HANDBOOK OF
MOISTURE DETERMINATION AND CONTROL

PRINCIPLES, TECHNIQUES, APPLICATIONS

VOLUME 3

HANDBOOK OF MOISTURE DETERMINATION AND CONTROL

PRINCIPLES, TECHNIQUES, APPLICATIONS

IN FOUR VOLUMES

by A. Pande
*Shriram Institute for Industrial Research
Delhi, India*

VOLUME 3

MARCEL DEKKER, INC.   New York

COPYRIGHT © 1975 by MARCEL DEKKER, INC. ALL RIGHTS RESERVED

Neither this book nor any part may be reproduced or transmitted in any form or by any means, electronic or mechanical, including photocopying, microfilming, and recording, or by any information storage and retrieval system, without permission in writing from the publisher.

MARCEL DEKKER, INC.

270 Madison Avenue, New York, New York 10016

LIBRARY OF CONGRESS CATALOG CARD NUMBER: 73-86820

ISBN: 0-8247-6186-3

Current printing (last digit):
10 9 8 7 6 5 4 3 2 1

PRINTED IN THE UNITED STATES OF AMERICA

Dedicated to the Memory of

Sir Shri Ram

Industrialist, Philanthropist, Educationist

and Founder of the

Shriram Institute for Industrial Research, Delhi

GENERAL PREFACE

The question of moisture determination and control is an important one in many scientific and industrial disciplines. The available information, however, is scattered through the literature in divers scientific journals. This monograph represents an attempt to create a basic source book to bring these various contributions together as a unit. It also includes for the first time many unpublished results obtained by the author and his associates during the course of their investigations. Many scientific disciplines (for instance chemistry, physics, statistics, and electronics) are covered and are synthesized in the techniques of moisture determination and control.

The four-volume monograph discusses the phenomena of moisture sorption, total versus 'free water' and 'bound water' effects, equilibrium relative humidity-moisture content relationship and describes most of the physicochemical methods and techniques with the latest developments in moisture determination and control in various materials such as food stuffs, cereals, grains, textiles, bagasse, pulp, paper and paper products, coal, coke, chemicals (even in the ppm range) and biological materials.

The monograph is broadly divided into two groups, i.e., methods and techniques and their applications. A certain amount of overlapping could not be avoided as specific applications had to be described to illustrate the effectiveness of a particular method or technique. These methods and techniques were developed independently by research workers in their own fields, often unaware of similar work in other fields. All of these have been compiled for the sake of completeness of the state of the art.

It is believed the value of the monograph has been enhanced to a large extent due to the indexing and bibliography provided in Volume 4 to facilitate the location and extraction of information. The latest pertinent references to the original publications and illustrations have been provided for further study of this fascinating field. It is earnestly hoped that the book will serve the purpose for which it has been compiled.

Any suggestion for the improvement of the monograph would be highly appreciated.

A. Pande

Shriram Institute
Delhi (INDIA)
September, 1974.

PREFACE TO VOLUME 3

This volume marks the beginning of the applications of the moisture determination and control methods and techniques to different materials such as natural and manmade fibers, wood, bagasse, paper and paper products, grains, cereals and other agricultural products, such as coffee, tea, tobacco and soils, sands and other similar materials, such as coal, coke and other granular materials. Before describing the applications of different methods of moisture determination and control, the theoretical and practical aspects of the phenomenon of moisture sorption in these materials have been outlined.

Discussion of the comparative merits and shortcomings of various methods and techniques as applied to these materials has been provided wherever possible, and tables showing these comparisons have been included.

Commercial instruments and equipments which are used in industrial processings for monitoring or controlling moisture contents of processed materials are briefly described for the sake of completeness of the subject matter. Inclusion or omission of the name of a particular company, however, does not signify any bias or prejudice to the company. It is quite likely many other commercial organizations have developed similar instruments and equipments.

The author gratefully acknowledges the cooperation received from scientists working in this field who provided him with the reprints of their publications. Data gleaned from these have been included in this monograph. The author is specially thankful to editors and publishers of the following journals who gave permission for the reproduction of illustrations from their publications. These are Journal Colloidal Science, American Chemical Society, Analytical Chemistry, Journal of National Bureau of Standards, Textile Praxis, Textile Research Journal, and British Journal of Applied Physics and Fuel.

The author is thankful to manufacturers of moisture measuring and controlling instruments and equipments who provided technical literature with illustrations. These are Electronic Automatic System, U.S.A., Fielden Electronics, U.K., Nucleonics, U.S.A., Minneapolis Honeywell Regulating Company, U.S.A., Electrodynamics Corporation, U.S.A., Marconi Instruments, U.K., Motomco Inc., U.S.A., Foxboro Moisture Measuring System, U.S.A., C.S.I.R.O. Regional Laboratory, Australia, Associated Electrical Industries, U.K., and British Concrete Industries Research Association, U.K.

The author is thankful to Dr. Charat Ram for granting permission to publish this monograph and dedicates it to Sir Shri Ram, founder of Shriram Institute. The author is also thankful to Dr. John Mitchell, Jr., Manager, Research and Development Division, E. I. du Pont de Nemours & Co., who kindly reviewed the manuscript and made very useful suggestions.

The author gratefully acknowledges the help given by Mr. Ved Parkash of N.P.L. for the survey of literature and illustrations and Mr. S. K. Ganeriwalla, Mr. R. N. Singh, and other colleagues for providing data on the applications of the S.R.I. moisture meter in different materials. Thanks are also due to Mr. P. M. Goel who typed the entire manuscript so skillfully.

Finally the author is very grateful to the publisher and his editorial staff who gave unflinching cooperation during the publication of this monograph.

Shriram Institute
Delhi (INDIA)
December, 1974

A. PANDE

CONTENTS

GENERAL PREFACE	v
PREFACE TO VOLUME 3	vii
CONTENTS OF OTHER VOLUMES	xi
CHAPTER IX. MOISTURE IN TEXTILES	**585**
1 Introduction	585
2 Moisture in Cotton	586
3 Sorption of Moisture by Man-Made Fibers	604
4 Moisture Sorption by Wool and Similar Fibers	611
5 Moisture Sorption by Jute and Similar Vegetable Fibers	644
References	650
CHAPTER X. MOISTURE IN BAGASSE, WOOD, AND PAPER	**657**
1 Moisture in Bagasse	657
2 Moisture in Beet Pulp	666
3 Moisture in Wood and Timber	669
4 Moisture in Wood Chips	685
5 Moisture in Paper and Paper Products	689
6 Conclusion	710
References	712
CHAPTER XI. MOISTURE IN FOODS AND ALLIED AGRICULTURAL PRODUCTS	**717**
1 Introduction	717
2 Sorption of Water by Cereal Grains	718
3 Moisture in Flour and Flour Products	743
4 Moisture in Sugar and Sugar Products	749

5	Moisture in Dried Fruits	759
6	Moisture in Dehydrated Foods	766
7	Moisture in Milk Products	771
8	Moisture in Raw Beverages	774
9	Moisture in Condiments and Spices	779
10	Moisture in Tobacco	779
	References	780
CHAPTER XII.	MOISTURE IN SOILS, SANDS, CONCRETE, AND SILICA AND SILICATES	789
1	Moisture in Soils	789
2	Moisture in Concrete and Similar Materials	853
3	Moisture in Silica and Silicates	864
	References	867

Cumulative Indexes will appear at the end of Volume 4.

CONTENTS OF OTHER VOLUMES

VOLUME 1

Chapter I   Water, Its Properties and Interaction with Hygroscopic Materials
Chapter II   Gravimetric Methods and Techniques
Chapter III   Azeotropic Distillation and Chromatographic Methods and Techniques
Chapter IV   Karl Fischer Method and Its Applications

VOLUME 2

Chapter V   Electrical and Electronic Methods
Chapter VI   Spectroscopic Methods and Techniques
Chapter VII   Nuclear Methods and Techniques
Chapter VIII   Automatic Control of Moisture

VOLUME 4

Chapter XIII   Moisture in Coals and Similar Materials
Chapter XIV   Moisture in Chemicals and Their End Products
Chapter XV   Moisture in Biological and Biochemical Materials

Chapter IX

MOISTURE IN TEXTILES

1 INTRODUCTION

The sorption of moisture by textiles and similar hygroscopic materials is of considerable interest to those engaged in the manufacture, processing, and finishing of these materials. A large number of papers [1-5] and reviews [6-11] have been published dealing with the theoretical and experimental aspects of the phenomenon of sorption of moisture in textiles, involving exchange of heat as well as other associated physicochemical effects. The relationship between moisture content and physicochemical properties of the fibrous materials has also been investigated by a number of investigators [12-15] and the effects of these properties on the performance characteristics and end uses have been evaluated.

In this chapter, the phenomenon of sorption of moisture in such fibrous materials as cotton, jute, wool, silk, and synthetic fibers is described, and the various methods of measurement of regains in them are discussed with a view to explain why a particular mode of approach, method, or technique has been found to be more successful for measuring moisture content of textile materials in bulk as well as during the industrial processing, i.e., sizing, dyeing, finishing, etc.

2 MOISTURE IN COTTON

2.1 Sorption of Moisture by Cotton

The general theory of sorption of moisture by hygroscopic materials has been described in the literature and is applicable to textile materials as well. In textile fibers moisture uptake depends on the configuration of the molecules in the amorphous regions. The water is absorbed in the amorphous parts of the fiber where it is uniformly distributed, and the moist fiber has the properties of a solution of water in cellulose. It is well known that the sorption is attributable to hydrogen bonding of water molecules to the available hydroxyl groups of the substrate, i.e., those in the amorphous regions and on the surfaces of the crystallites. Measurements of the energy changes accompanying sorption suggest that the initial strong sorption occurring at the lowest values of the relative vapor pressure results from the linking of one molecule of water to a pair of suitably placed hydroxyl groups of the cellulose, while at later stages combination to one hydroxyl group and to the oxygen atoms occurs, while finally water molecules are able to condense only on already heavily covered surfaces and the thermodynamics of the process approaches that of condensation in bulk.

The sorption of water by cotton was first accurately measured by Urquhart and Williams [16] who found the saturation regain for the soda-boiled cotton to be between 22.42 and 22.97% and the normal regain to be 8.5%. The moisture regain of cotton in equilibrium with an atmosphere was found to depend on the condition of the cotton prior to its exposure to the atmosphere. It was found that the regain values were different during the processes of absorption and desorption. This is due to the well-known phenomenon of hysteresis

2 MOISTURE IN COTTON

in which the absorption curve lies lower than the desorption curve (Fig. 9.1). The moisture regain depends also on the previous heat treatment; i.e., previous heating to a high temperature reduces the capacity of the material to absorb water. Soda-boiled cotton produces a greater hysteresis effect than fresh cotton but takes up less water when absorbing moisture. It was found, however, that the hysteresis of cotton does not extend to zero humidity.

The relation between the moisture content of textile fabrics and the relative humidity of the atmosphere has been examined by Müller [17], who derived a formula correlating the two constants for calculating the moisture content at different humidities. He also reported the phenomenon of hysteresis, i.e., the absorption curve lies lower than the desorption curve.

It has been observed by Richter et al. [13] that sorption of water is affected more by changes in fiber properties at 95 to 100% relative humidity than at lower relative humidities. The sorption levels are shown to be dependent not only on chemical history but also on the condition and the extent of water removed before testing. Sorption of organic vapors at high relative humidities depends on the vapor in question and on the extent to which the fiber has been

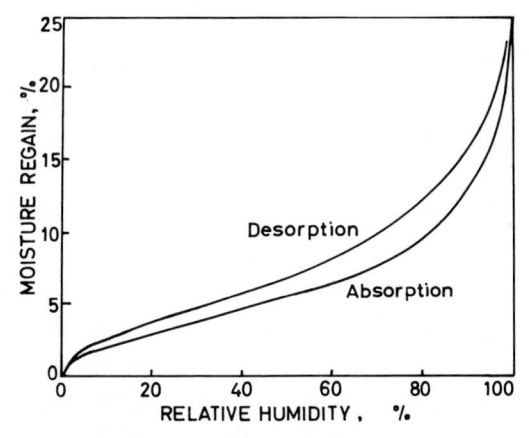

FIG. 9.1. Graph showing comparative hysteresis effects of pure cellulose and scoured cotton.

swollen at the time of exposure to the vapor. Both rate and quantum of sorption of such vapors depend on the nature of the cellulose; however, neither parallels the sorption of water by the same cellulose.

Iyengar and Prakash [18] have determined the standard regain values for Indian cellulosic textiles including raw cotton, cotton yarn, and cotton fabrics after conditioning them at a relative humidity of 65 ± 2% and a temperature of 80 ± 2°F. From the detailed statistical analysis carried out on individual test results of raw cotton, the conclusion was drawn by Iyengar and Prakash [18] that mean moisture regain for raw cotton before cleaning was 8.35%, which agrees fairly well with the 8.5% normally used for raw cotton. However, after cleaning with Shirley Analyzer, the mean regain was reduced to 7.82%. The effect of cleaning is found to reduce moisture regain percentage in all the three categories of cotton (coarse, medium and fine) the effect being maximum in the category of fine cotton. The variance of moisture regain percentage between cottons was found to be significant for uncleaned cotton, while it became insignificant after cleaning, thus showing that the trash content is responsible for the variability observed between cottons.

In the case of yarns, it was observed that the range of moisture regain lies between 7.11% (for coarse, dyed yarns) and 8.00% (for coarse, gray yarns), which is in close agreement with that given in the Textile World Yearbook and Catalogue [19], viz., 7 to 8%. It was further observed that the mean moisture regains were 7.70, 7.72, 7.63, and 7.58% for the yarn types, coarse, medium, fine, and superfine, respectively. Similarly the mean moisture regains of 7.81, 7.50, 7.44, and 7.72%, respectively, were found for the treatments gray, bleached, dyed, and ply. From the detailed nonorthogonal statistical analysis of the significance of these differences, it was concluded that the effect of fineness of count is negligible on moisture regain percentage, while the effect of treatment (gray, bleached, dyed, and ply) is found to be comparatively higher although statistically nonsignificant. The effect of treatment and count of yarn and moisture regain are shown in Fig. 9.2.

The total range of variation in moisture regain percentage in the case of fabrics was found to lie between 6.76% (in the case of dyed long cloth) to 8.86% (in the case of gray coating), which also agrees very closely with the range 6.5 to 8.5% given in Textile World Yearbook and Catalogue [19] for the case of fabrics. It was further observed that there are some variations between different types of fabrics and between different treatments both in the sized and desized states of the fabrics. The moisture regain ranges from 7.36% for long cloth to 8.08% for the corresponding fabrics after desizing. Similarly, it ranges from 7.23% for the dyed fabrics to 8.10% for the printed in the case of gray fabrics and from 7.34 to 8.19% for the printed in the case of raw fabrics and from 7.34 to 8.19% for the corresponding samples in the desized state. It was observed that the structure of the fabrics was found to have practically no effect on moisture regain percentage, which corroborates similar conclusions drawn with regard to the effects of fineness of cotton and of cotton yarn. Perhaps the effects of fineness of cotton, count of yarn, and structure of cloth are restricted to only the rate at which moisture absorption takes place, but final moisture

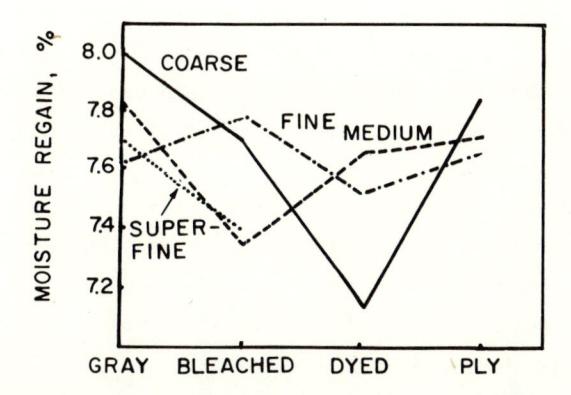

FIG. 9.2. Graphs showing the effects of treatment and count of yarn on moisture regain.

regain percentage may not be affected by them provided sufficient time is allowed for the sample to attain equilibrium.

Effect of Temperature. Urquhart and Williams [20] have carried out precise experiments dealing with the absorption of water by cotton at all humidities and at temperatures up to 110°C; these data are, therefore, of practical significance not only to the testing of cotton materials but also to the control of the conditions prevailing in processings of textile yarns and fabrics. Their results show that when the relative humidity is constant at a value of less than 80%, an increase of temperature from 100 to 110°C decreases the moisture regain of the cotton. When, however, the relative humidity is constant at more than 80%, a similar decrease in moisture regain occurs in the temperature range of 10 to 50°C, while from 60 to 110°C the moisture regain may actually increase with rise of temperature. It appears, therefore, that the increase of moisture regain, under constant high humidities, which begins at about 60°C, is due to swelling of the material and the **consequent** exposure of greater surface. Any process which depends for its success on the swollen state of the cotton will consequently become easier above 60°C and at high humidities.

From a series of experiments, Wiegerink [21] concluded that the logarithm of regain values at a given relative humidity is linearly related to the reciprocal of the corresponding absolute temperature. The effect of temperature on moisture regain was found to be greater at temperatures above 93°C. The uptake of water is an exothermic reaction, and therefore the amount of water taken up at constant partial vapor pressure decreases with increasing temperature. Figure 9.3 shows a series of isotherms for the absorption of water by cotton cellulose at temperatures in the range of 10 to 50°C and provides experimental confirmation of the theoretical deductions. This figure also shows that in the temperature range 50 to 100°C, the same effect is found at humidities below about 85% relative humidity, but at higher humidities the curves cross one another, indicating that in

2 MOISTURE IN COTTON

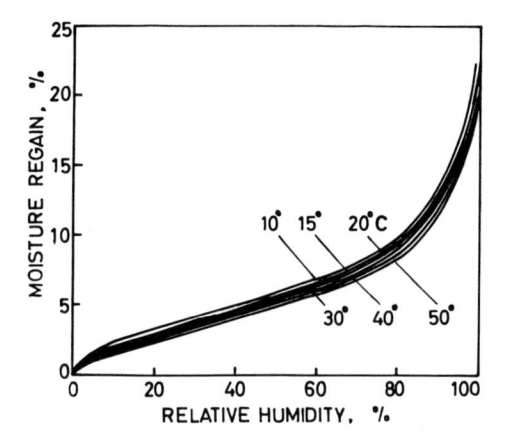

FIG. 9.3. Absorption curves of water by cellulose at different temperatures.

this region the amount of water taken up at constant humidity increases with increasing temperature. For an explanation of this phenomenon, the reader is referred to the original publication of Wiegerink [21].

2.2 Effect of Chemical Modifications

When cotton and rayon are treated with urea and melamine formaldehyde resins, their moisture pickup drops with the increase in resin content. However, even the highest concentration of resins used indicates that less than one hydroxyl group reacts per glucose unit. On this basis it is easy to see that the reduction in moisture pickup roughly corresponds to the effect of blocking off one hydroxyl group. Most probably one hydroxyl group reacts with the methylol compound which affects its moisture pickup and regain values. These crosslinking agents effectively prop open the fiber structure, and the degree of openness is dependent on the extent of fiber swelling at the time of crosslinking. Moisture regain of the crosslinked cotton is, therefore, more dependent on the state of fiber swelling at the time of crosslinking than on the extent of crosslinking [22-23].

Figure 9.4 shows isotherms for cotton, secondary cellulose acetate, and cellulose triacetate fibers, and illustrates how the initial bulge in each curve, corresponding to the strong sorption, tends to disappear as the number of hydroxyl groups in the molecule decreases. These commercial materials are, however, relatively highly substituted, and it is more interesting to consider the relative sorptions of samples of different degrees of substitution (DS). Results of such studies are shown in Fig. 9.5. Curve 1 shows the moisture sorption at 65% relative humidity of methyl cellulose prepared from a regenerated cellulose; the sorption rises to a maximum from the relatively high value of the regenerated cellulose itself, and falls off linearly as the DS increases, falling more rapidly as the DS approaches 3, and at this stage the contour of the molecule perhaps becomes sufficiently regular to permit crystallization. Curve 2 is for cellulose acetate and is of the same general shape, but starts at the lower value characteristics of cotton and rises to a higher value because of the greater hygroscopicity of the acetyl groups as compared with the methyl. On this basis, the intermediate linear portion may be regarded as representing the variation with DS of the moisture sorption of completely amorphous material, and the extrapolation of this line to a DS corresponding to zero should give the sorption value for completely amorphous cellulose, and to a DS of 3 for the completely amorphous derivative. The value obtained for cellulose is the same, as it should be, for both derivatives, and is about 17%. The fraction of the total hydroxyl groups in the cellulose which are accessible to water vapor can be determined by dividing by 17 the sorption value of the cellulose at 65% relative humidity.

Howsman [24] has shown that moisture regain of cellulose fibers is related to the average degree of lateral order. Since cellulose crosslinked in a swollen condition is stabilized in that state, its moisture regain and water of imbibition are higher than those of untreated cellulose. As the degree of stabilization increases, i.e., as the amount of the reagent combined becomes higher, the

2 MOISTURE IN COTTON

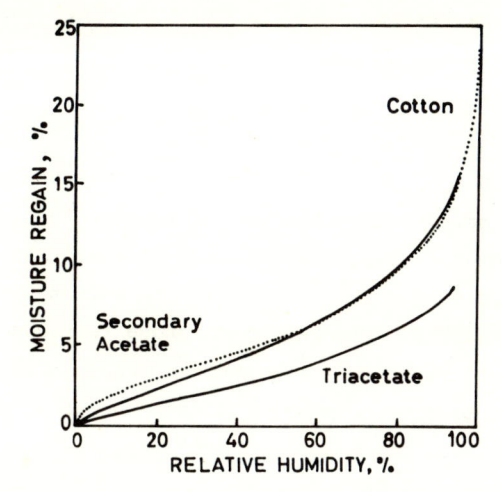

FIG. 9.4. Sorption curves for cotton, secondary cellulose acetate, and cellulose triacetate.

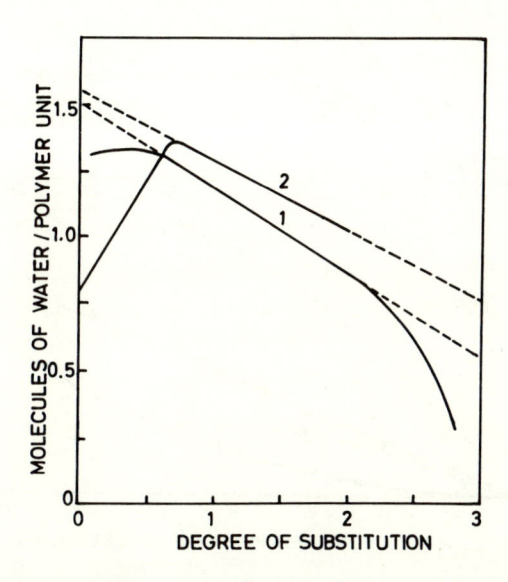

FIG. 9.5. Relative sorption curves of samples of methyl cellulose and cellulose acetate having different degrees of substitution.

moisture regain increases. This is shown in Fig. 9.6, wherein the percentage improvement in moisture regain over the control is plotted against the millimoles of different chemical compounds combined per 100 g of cellulose.

It was found by Blouin and Arthur [25] that the moisture regain of cellulose is not greatly affected by irradiation with gamma rays. The changes which do occur, however, are unusual. The regain values start to decrease at 5×10^6 R, reach a minimum at 5×10^7 R, and start to rise at 10^8 R. Changes in moisture regain are usually thought to reflect changes in the submicroscopic structure of the cellulose if the treatment has not altered the chemical composition of the fibers.

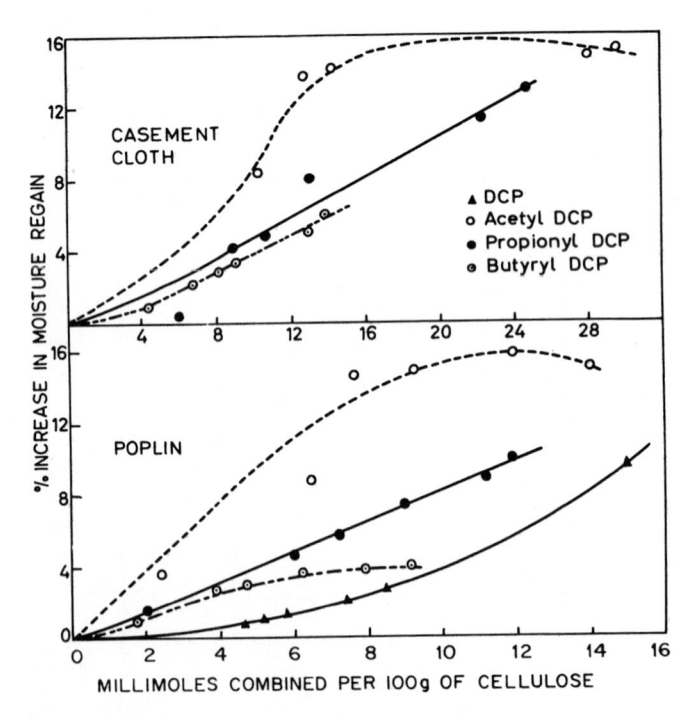

FIG. 9.6. Percent increase in moisture regain with increasing moles of esters combined.

2.3 Effect of Crosslinking

A comparison of the absorption isotherms (Fig. 9.7) of untreated, dry crosslinked, and wet crosslinked cottons shows that the dry treatment reduces the water absorption and the wet treatment increases it. This type of behavior was observed previously for similarly treated samples compared at a fixed relative humidity, usually of 65%. It is believed that dry-state treatment introduces covalent crosslinks between cellulose chains which might be expected to reduce swelling. However, in the case of the dry crosslinked rayon fabric, no reduction in swelling occurs, and at high humidities swelling is actually increased. This behavior is not understood, but it is suggested that the crosslink density in the amorphous portion of rayon is less than that in cotton, and that at higher crosslink densities in rayon a reduction in swelling would occur. The effect of wet-state crosslinking is to increase the water absorption of cotton, as shown in Fig. 9.7. The wet crosslinking process probably results in a fiber structure more accessible to water by introducing a higher portion of intramolecular crosslinks to be fixed between disordered cellulose molecules in the outer fringes of crystalline regions which would normally recrystallize during drying. The result is a more open cellulose structure into which water can penetrate more easily.

It has been observed by Reeves et al. [23] that moisture regain of the crosslinked cellulose depends on the conditions under which they are crosslinked, and it may be observed from the data of Gill and Steele [26] that the moisture regain of the cellulose crosslinked in the swollen state has a higher value than that of the untreated cotton cellulose at 65% relative humidity. Moisture sorption isotherms of cellulose treated with various amounts of formaldehyde at different degrees of swelling have been obtained by Pande and co-workers [27] in the range of zero to 90% relative humidity and at 25, 35, and 45°C. It was observed that the maximum moisture sorption is obtained corresponding to a formaldehyde add-on of 0.49% at 16%

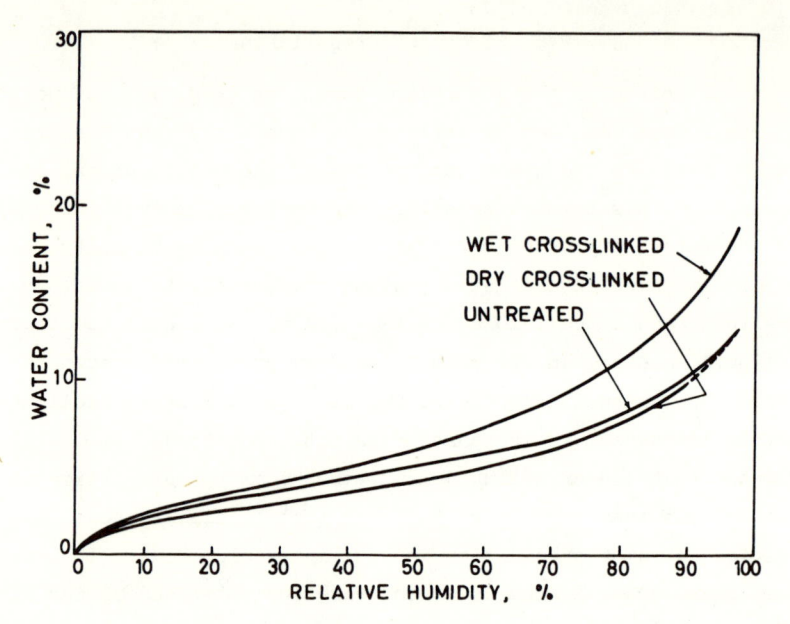

FIG. 9.7. Absorption isotherms of untreated, dry crosslinked and wet crosslinked cotton at 25°C.

swelling. In cellulose acetate fibers, which are derived from cellulose by the replacement of most of the hydroxyl groups by acetyl groups, the side groups are inert and do not attract water strongly. Consequently, the moisture absorption of acetate is less than that of unmodified cellulose.

2.2 Methods of Moisture Measurement

A number of methods of moisture measurement in cotton textiles are available and are described in the literature. These methods can be broadly classified as physical and chemical methods. The chemical methods consist of oven-drying techniques, resulting in the development of thermobalances, absorption techniques, azeotropic distillation techniques, the Karl Fischer and other titration methods and improvements thereon. The physical methods comprise electrical or electronic methods, infrared spectroscopy, neutron scattering, mass spectroscopic techniques and nuclear magnetic resonance devices.

These methods have been described and discussed by the author in a series of publications [6-9].

The general methods of determining moisture (content) usually involve gravimetric techniques, and those for textile materials are no exception. Standard oven drying is the oldest and most fundamental method. Specifications for standard oven-drying procedures are provided by Tappi [28] and the American Society for Testing Materials [29]. Standard oven drying is perhaps most widely used but involves a lengthy procedure, does not differentiate between moisture and other volatile products, and careful experimentation is necessary for getting accurate results. Because of the extremely hygroscopic nature of cellulose, the dried sample must be enclosed almost airtight to prevent moisture absorption while weighing; thermally unstable products cannot be heated in this way as the fibers tend to decompose. Vacuum techniques, however, can be applied with advantage. The sample of the textile material to be analyzed by oven drying is weighed to a constant weight at a temperature of 105 to 110°C and the loss in weight is assumed to be the moisture content of the material. No attempt is made to collect the moisture involved in this process, as distinct from most other techniques (distillation, for example) in which the water is collected separately before making the necessary measurements. Factors involved in oven-drying procedures have been discussed by Mitchell [30], who has shown experimentally that careful standardization is essential for getting reliable and repeatable results. In the hands of an experienced technician, however, the measurements should be reproducible to within ±0.2% moisture content.

Attempts have been made to develop more refined drying equipment or thermobalances in place of conventional oven-drying techniques. Ghöde [31] has devised an apparatus of this kind for the determination of moisture in cellulose derivatives, consisting of an electrically heated cabinet with built-in automatic temperature regulator, air circulator, and projection reading balance. With this apparatus, he could drive off moisture from cellulose, wool, pulp, and paper in a fraction of the time required in the case of normal oven-drying methods. Such thermobalances have been described in an excellent book on thermogravimetry by Duval [32]. In certain cases it is undesirable to have to adjust the amount of material to a precise value or the sample may be of a material such that its weight is not readily adjustable (e.g., a length of textile yarn). A direct reading balance which does not require a fixed initial weight of material would frequently be of advantage. Such balances have been described by Barker and Hedges [33] and by Downes [34].

Brock [35] has developed a novel and quick method of moisture measurement by using infrared radiation as a source of energy to evaporate and drive off the moisture content of the materials. He found that infrared radiation was of great advantage in bone drying quickly (within 5 min) even a pile of fibers $\frac{1}{2}$-in. thick without raising its temperature above 70°C. Based on these observations, he developed a moisture meter for cotton and similar hygroscopic materials utilizing infrared radiation. He obtained an accuracy of 0.2%, which is considered to cover the testing requirements of most industrial laboratories. A similar but cheaper equipment for measuring moisture content of cotton, wool, and staple fibers has been developed by Hummel [36] which takes 30 to 60 min to measure the moisture content. In Table 9.1 the values of moisture content of some textile materials as obtained by this meter are compared with those obtained by the conditioning oven. From this table it will be observed that the values obtained are higher. This is due to the fact that in the conditioning oven, the hot air is forced through the sample, which dries the sample much more uniformly and thoroughly.

Absorption techniques (see Chap. II for a detailed description) are frequently used in the laboratories for the determination of moisture contents. They are refinements of oven-drying methods, for the water vapor which is evolved at elevated temperatures is carried along an inert-gas stream to a vessel containing one of the suitable absorbents. Certain chemicals such as phosphorus pentoxide (P_2O_5), anhydrous magnesium perchlorate (Dehydrite), calcium sulfate

TABLE 9.1

Comparison Between Test Results Obtained with
Infrared Radiation Drying and the Conditioning Oven

Material	Regain values, as obtained on infrared moisture meter (%)	Regain values, from conditioning oven (%)	Difference
Wool (conditioned)	12.3	13.2	0.9
Cotton (loose stock)	6.9	8.0	1.1
Staple fibers (loose stock)	12.2	13.8	1.6
Staple fiber yarn	10.1	10.9	0.8
Wool yarn	13.3	13.5	0.2
Cotton fabric	7.8	8.6	0.8

(Drierite), calcium chloride, and barium oxide have great affinity for water and associate very readily with the ambient moisture. They are, therefore, quite often used as dehydrants in moisture determinations in conjunction with either dry hot air or under vacuum. Kuntzal [37] has evolved a method for measuring the moisture content of loosely packed solids by circulating hot dry air and absorbing the effluent vapors in tared drying tubes. This method can be adopted for cotton textile materials as well.

The Karl Fischer method has been successfully employed for the moisture determination of cotton and other fibers. The ease with which the moisture in cotton can be titrated with the Karl Fischer reagent was first reported by Zimmerman [38]. Although no experimental details were given, values of 6.53 and 6.65% moisture content were found by titration as opposed to 5.66 and 6.40% weight loss by oven drying after 5 hr at 105°C. Similarly, somewhat higher results for cotton linters by titration were reported by Mitchell [39]. This cold extraction procedure was found to be quite applicable for the determination of moisture content of cotton textiles. Keating and Scott [40] found that a 10-min contact period with a sample containing 50 to 25 g of water with 100 ml of methanol was enough to extract all the moisture from cotton yarn or fabric. In carrying out the analysis, no attempt was made to condition the samples in an atmosphere of constant temperature or humidity prior to the determinations. This volumetric method is equally effective whether the cotton is in yarn or fabric form.

The Karl Fischer method takes only one-fifth of the time required for the drying process. A significant advantage of the Karl Fischer method exists for those substances which lose their constituents during drying, and more so for the cases where highest accuracy is the goal. In conclusion, it can be stated that the Karl Fischer method has several advantages over oven-drying methods. It is rapid and on a routine basis it requires 10 to 15 min/sample. The value of moisture content as determined by this method represents true water content and not a loss in weight as is the case in the oven-drying methods. The method is calibrated against a weighed amount of water, hence the results obtained are quite accurate. A precision of ±0.1% is attained with two portions tested, and seven portions will give a precision of ±0.05%. Increasing the size of the portion in the Karl Fischer determination also increases the precision. The use of formamide in the Fischer reagent assures a complete extraction of water with a minimum of heating. Interfering substances give little or no trouble.

The textile industry has long recognized the need for a rapid and accurate method of determining the moisture content of cotton and other textile materials. Research has continuously provided a wealth of knowledge on this subject and (as a result) a number of useful and accurate instruments have been developed for the measurement of moisture content in bulk as well as in continuous processes. Most of the instruments developed so far that operate with speed and accuracy are of the electrical type which utilize the electrical properties of water, i.e., conductivity and dielectric. The electrical instruments can be divided into two general classes, viz., the conductivity or resistance type, in which the electrical resistance of the sample is correlated with the percentage of moisture, and the capacitance type, in which the dielectric constant is related with moisture content. A detailed description of the electrical or electronic moisture meters is given in the literature [41] and in Chap. V. Some of the moisture meters developed specifically for textiles and their applications to these materials are briefly discussed here.

The conductivity principle has been used by Jones [42] in developing the Shirley moisture meter, which is one of the most accurate electrical moisture meters for cotton and similar hygroscopic materials. The electrical system of the Shirley moisture meter consists of a commercial-type resistance meter with built-in voltage-stabilizing features, connecting shielded cables, and two flat plate electrodes. The electrodes are 5 in. in diameter, one movable and the other stationary. With the electrodes arranged in this manner, the current passes through the whole sample mass. It was found that this method produces much better results than an arrangement in which the electrodes are in the same plane with the current flowing in a lateral direction. This latter method gives an indication of surface moisture only, and is not fully representative of a sample with uniform moisture content, e.g., well-conditioned samples measured under a standard atmospheric condition.

The sample must be kept under uniform compression during the test for good results. During the course of their investigation of moisture measurement of lint cotton, seed cotton, etc., it was found by Reddick et al. [43] that a surface force of 40 psi produces the best results. To produce and maintain the required compression force, hydraulic power was selected by them because of the excellent control possibility and because it is relatively inexpensive.

Fielden [44] has developed a capacitance-type moisture meter for cotton fabrics. This meter is capable of giving reliable readings of moisture content in increments of 0.05% or less down to complete dryness with an accuracy of ±1% in the range of dry to 20% moisture content. This instrument has been applied successfully as a control (monitoring) device on sizing machines in the cotton textile industry. Based on the principle of "off the resonance," Pande [45] has developed a capacitance-type moisture meter which has been found applicable to cotton textile materials in bulk as well as for use as a monitoring device on a sizing machine or tenter.

Electronic and electric meters in general show values about 0.5% higher than those given by oven-drying methods, but the capacity types are accurate up to ±0.2% in the range of 5 to 20%; these results were arrived at by Whitewell, Bowen, and Toner [4-5] who compared the performance of electrical meters with standard oven-drying methods. For highest accuracy, the meter must be calibrated not only for the type of fiber but for each type of dye or finish.

It was found by a committee of Textile Technologists [46] in the United States that bleaching, mercerizing, and scouring involve only minor changes in the substrate, and these effects are considered negligible as far as moisture is concerned. Any electrolyte (and many dyes are electrolytes) will affect meter readings, but whether or not seriously will depend on the tolerances required, the electrolyte concentration, and presumably also on the electrical parameter being measured. Results as determined by these experts on cotton and viscose fabrics [46] are reported as instrument scale readings at moisture regains based on moisture determinations by standard oven-drying methods. This method of presentation has value because in effect it gives separate meter calibration data for base fabrics and treated fabrics.

The effect of dyeing and resin treatments on electric moisture meter readings obtained with both cotton and rayon at several moisture regains is presented in Tables 9.2 and 9.3. In general, at a particular regain, the treatments appreciably alter meter readings as compared with those for the untreated fabrics. The dye treatments have the effect of giving higher readings on three commercial electrical meters examined. The resin treatments, on the other hand, resulted in lower meter readings.

TABLE 9.2

Data Showing the Effect of Different Dyes on Meter Readings as Compared with Oven-Drying Values[a]

Fabric	Oven moisture regain (%)	Undyed base fabric	Fabric dyed with				Undyed resin-treated fabric
			Sulfur dyes	Direct dyes	Vat dyes	Naphthol dyes	
Bleached cotton	6	16.0	-	-	-	-	8.8
	7	19.6	-	-	32.6	31.6	10.4
	8	23.4	39.4	44.8	40.6	39.6	11.8
	9	27.0	47.6	49.0	49.0	47.6	-
	10	30.8	50	50	50	50	-
Spun rayon	13	22.0	-	-	-	25.0	17.6
	14	23.4	24.6	27.4	27.4	-	19.0
	15	25.4	28.2	-	31.6	33.6	-
	16	28.2	-	35.0	-	-	-
	18	33.8	38.8	43.0	44.8	46.6	21.6

[a] The meter readings given here do not indicate directly moisture content values. These are determined graphically from the scale readings.

A recent paper by Watt and Leeder [47] gives comprehensive references to the existing literature on the subject. The overall impression gained from published work is that the effect of dyes on the equilibrium regain around 60% relative humidity is generally small, but the dyed material may range in regain from 1% above to 2% below that of undyed material. It was found by Baird and Fitz [48] that the shrink-proofing treatment on undyed top caused on average a decrease of regain of about 0.4%. However, on dyed tops, the same treatment did not cause significant changes in the regain (about 0.05% in one case and +0.2% in the other). The dyed and shrink-proofed tops were on average about 0.2% lower in regain than the untreated, undyed tops.

TABLE 9.3

Data Showing the Effect of Printing and Resin Treatment on Meter Readings as Compared with Oven-Drying Values[a]

Oven moisture regain (%)	Bleached white cotton base cloth	Printed only	Printed and treated with 6% urea formaldehyde resin	
			Uncured	Cured
6	7.0	-	-	8.8
6	0	-	-	4.5
6	12.8	45.0	-	38.0
7	8.0	9.7	10.0	10.0
7	4.5	8.0	12.0	16.6
7	16.4	46.2	44.4	48.0
8	9.7	10.0	10.0	10.0
8	44.0	12.5	48.0	0.5
8	20.7	47.6	50.0	50.0
9	10.0	10.0	10.0	10.0
9	48.0	14.0	3.0	5.0
9	27.6	50.0	50.0	50.0
10	10.0	10.0	10.0	10.0
10	14.0	-	-	-
10	34.2	50.0	50.0	50.0

[a] The meter readings given here do not indicate directly moisture content values, these are in arbitrary units; the moisture content values are determined graphically from the scale readings.

3 SORPTION OF MOISTURE BY MAN-MADE FIBERS

3.1 Sorption by Rayons

The sorption of moisture (water) by rayons has been investigated by Urquhart and Eckersall [49], and as a result of these investigations they have found that the regenerated cellulose rayons give absorption and desorption curves similar to those given by cotton and

3 MOISTURE IN MAN-MADE FIBERS

mercerized cotton with like hysteresis. Water absorption isotherms at 25°C of rayon preswollen in NaOH solution and treated with DMEU (dimethylol ethylene urea) are shown in Fig. 9.8 from which it will be observed that the pretreatment in NaOH has an appreciable effect on its moisture sorption characteristics. The hygroscopic properties of viscose, cuprammonium, nitro, and acetate silks, and of viscose and cuprammonium silks of different denier numbers have been studied by Biltz [50]. It was found that viscose and cuprammonium "silks" have the same average moisture capacities and that these are independent of their denier numbers. Nitrosilk has a higher average moisture capacity, while that of acetate silk is very low. These results are in agreement with the observations that denitrated nitrosilk and viscose and cuprammonium silks are cellulose hydrates, while acetate silk is an ester having a low degree of hydration. The equilibrium moisture content of acetylated silk fibroin is slightly lower than that of nonacetylated fibroin and varies with the degree of acetylation. Casein fibers have regains very close to those of wool at the same humidity but show rather a large difference between the absorption and desorption values.

3.2 Sorption by Synthetic Fibers

The sorption of moisture changes the properties of synthetic fibers. It causes swelling to occur, thus altering the dimensions of the fiber, and this, in turn, effects changes in the size, shape, stiffness, and permeability of yarns and fabrics. The mechanical properties and the frictional properties also are altered, as is the behavior of the fibers in processing and in use. The electrical properties of most fibers are also considerably changed and static is much less likely to occur in damp conditions.

The synthetic fibers have low regains. According to Hutton and Gartside [51] silk has a regain intermediate between cotton and wool; silk gum has a high regain, and the degumming causes a reduction in regain at 65% relative humidity from 10.65 to 9.9%. The cellulosic fibers are hydrophilic and the synthetic fibers are water insensitive,

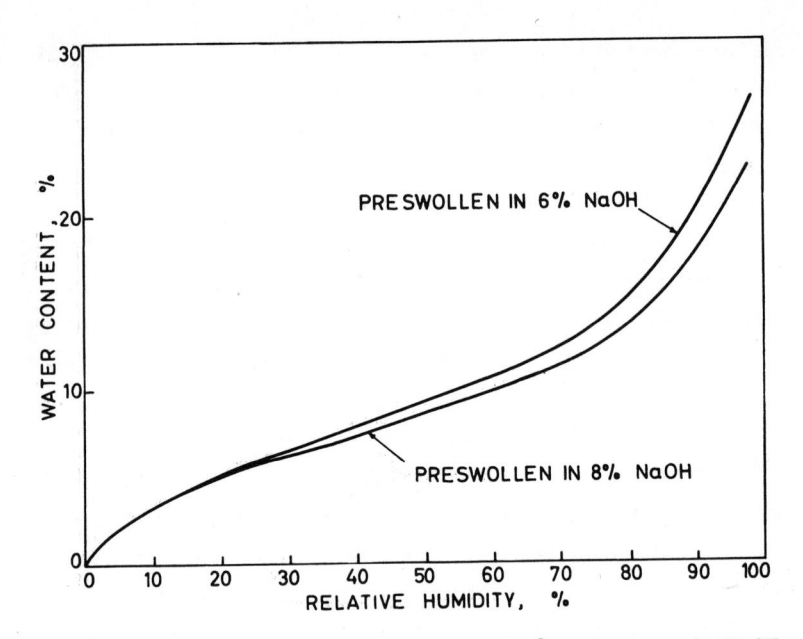

FIG. 9.8. Moisture sorption isotherms at 25°C of rayon in NaOH treated with DMEU.

i.e., hydrophobic. In general the regains of synthetic fibers such as nylon and polyester are about one-third to one-half of those of cellulosic materials. Some investigators have suggested that there is no hysteresis curve in the case of nylon, but Hutton and Gartside [52] showed that a small hysteresis definitely exists. At 80% relative humidity they obtained regain values of 5.48% in absorption and 5.64% in desorption. Forward and Smith [53] have shown that the moisture sorption of undrawn nylon yarn can be appreciably altered by chemical treatment. Armstrong and Walsh [54] have studied the physicochemical properties of textile fibers modified by the radiation-induced graft polymerization. Moisture regains were also measured on these fibers. It was found that grafted acetate shows a marked decrease in regain. The regains of nylon, however, were not significantly changed.

Untreated, unmodified fibers of polypropylene have a slippery, greasy, and cold feeling. The ability of such fibers to retain

3 MOISTURE IN MAN-MADE FIBERS

moisture is practically nil. Consequently, fabrics made from such materials rate very low in statistic behavior and in moisture regain. The imperviousness of polypropylene to water results from the highly crystalline structure of the polymer and from the absence of the waterbinding polar groups found in most other fibers. However, the moisture regain of these fibers increases after chemical and thermal modifications. By grafting CMS to the fibers and then treating them with pyridine, marked improvements are made in all these properties. The improvements effected are enough to make the fibers just as acceptable as other typical synthetics. Modified polypropylene fibers have a moisture regain (16% CMS graft, pyridine treated) of 3.88%, which is higher than the ratings accorded to Dynel (0.3), Dacron polyester fiber (0.4), Orlon acrylic fiber (0.9), Acrilan acrylic fiber (1.7), and nylon (3.8), but lower than the ratings for acetate, cotton (8.5), viscose yarns, and wool.

3.3 Methods of Moisture Measurement

Angster and Wandel [55] have applied the Karl Fischer technique for the determination of moisture in man-made fibers and have given a very interesting explanation for the high and low values obtained by this method. They have found that the weight of the sample should be so selected that the fiber materials should contain 100-180 mg water with especially low-water-containing fibers (polyvinyl chloride and polyester fibers, for example). The fibers must remain covered with methanol during the extraction process. Table 9.4 gives the average moisture content values of some of the fibers at 65% relative humidity according to the observations of Angster and Wandel [55]. The Karl Fischer method has been applied by Achwal [15] for the estimation of moisture in textile materials by determining the necessary period of extraction of water from cotton and viscose. He found that in all cases a constant value of moisture regain was obtained for periods of extraction of between 30 and 60 min.

TABLE 9.4

Average Moisture Regain Values for a Range of Man-Made Fibers Obtained by the Karl Fischer Method

Fiber	Moisture regain (%)
Viscose and cuprammonium fibers	12-13
Acetate fibers	6
Wool and casein fibers	13-14
Natural silk	10
Polyamide fibers (nylon, Perlon, etc.)	4
Polyacrylonitrile fibers (Orlon, Dralon, etc.)	1
Polyester fibers (Terylene, Diolen, etc.)	0.6
Polyvinyl chloride fibers	0.2

When one compares the values of moisture content as obtained by using the Karl Fischer method with those obtained by the usual oven-drying techniques, appreciable disagreement is observed. These differences are either positive or negative. Positive values (water content according to the Karl Fischer method yielding higher values than by the oven-drying process) arise in those fibers which have hydrophilic character, such as wool, natural silk, viscose, copper fibers, cotton, and flax. The explanation for the discrepancy is that water is combined with these fibers by strong secondary valencies and the capillary forces to such an extent that water cannot be fully eliminated at 105-110°C. Negative values (water content according to the Karl Fischer method is lower than that by the drying process) are obtained with fibers with low water content, such as, for example, polyvinyl chloride, polyester polyamide, and acetate fibers. These are the fibers having more of a hydrophobic character. In these fibers strong secondary valencies and strong capillary forces are lacking, consequently water is removed completely on drying. The differences are of course very negligible. Large divergences are noticed only when the fibers contain impurities which evaporate partially on drying. In the case of hydrophobic fibers, the difference in the dried and undried fibers is insignificant and this difference becomes significant only when fibers contain impurities which are driven off when the drying process takes place.

Infrared spectroscopy in the near-infrared region has been used for measuring the moisture content of natural and man-made fibers. This method is highly sensitive and is capable of giving very precise and absolute measurements. It has been used for estimating residual moisture in fibers by Elliot et al. [56] who obtained a strong absorption band at 5150 cm^{-1} due to moisture uptake by regenerated protein fibers. Jones [57] has also used this technique for accurate determination of moisture content in polymers (acrylic resins) in the range 0.2%. An accuracy of 0.05% or better was obtained, which is rather impossible to achieve with electrical methods. The accuracy of measurement of infrared spectroscopy is comparable to that possible by using the Karl Fischer method though the former requires calibration.

A rapid procedure for the simultaneous determination of caprolactum monomer and moisture in 6-nylon flake and yarn is described by Schenker [58]. They are separated from the sample by vacuum extraction at 200°C, the monomer and part of the moisture being condensed in a water-ice cold trap. The monomer is determined by refractive index in an aqueous dilution of the condensate, while moisture is obtained by difference, as the total sample weight loss represents monomer plus moisture. The procedure requires less than 3 hr as compared to the 24-hr water-extraction procedure for the determination of total water-extractable material, and when used in conjunction with the water-extraction procedure, it furnishes an estimate of the dimer content plus any low molecular weight water-soluble linear molecules present in the polymer. Haslam and Claser [59] have described a method for the determination of water in nylon which is based on heating the nylon in vacuo at 260°C for a definite period of time, collecting in a cold trap the water evolved, and subsequently determining it with Karl Fischer reagent.

Observations of moisture regain at room temperature have been conducted on GR-S test pieces by Lyons et al. [60] in which cotton and rayon tire cords had been vulcanized. These pieces consisting of thin pads containing the tire cord as flat fabric, and of blocks 1 in. thick, containing bundles of cords, were stored in jars at 100% relative humidity for periods extending up to 1 year. At the end of the humid storage period for the thin pads, they were removed to desiccating jars containing calcium chloride. Observations of moisture absorbed or desorbed were made by weighing the test pieces at appropriate intervals during the storage periods. GR-S test pieces containing no fabric were carried through the experiments as controls, to provide data on moisture in the rubber of the actual fabric-containing test pieces. From the results inferences are drawn concerning the percentage of moisture in the fabrics of actual tires under various humidity conditions of operation.

A comparative study of the surface resistivity of nylon has been made by Walker [61] and it has been compared with those of cotton and wool. The surface resistivity of fabrics made of nylon is found to be of the same order of magnitude as that of wool but has a higher value as compared with that of cotton. The resistivity of nylon at 0°C and 4% relative humidity is found to be 5×10^5 ohms/cm^2, and at 25°C and 50% relative humidity, it is 2×10^{-14} ohms/cm^2.

The measurement of the resistivity of nylon can be employed for estimating the amount of residual water present in nylon, which is an important factor in the plant control of the finished product, and, although there are very few references in the literature to the determination of water in nylon, various methods have been used in practice. These methods have not been found wholly satisfactory. In general they depend on estimating the loss sustained by nylon under various heat treatments by weighing the nylon before and after heating. This loss need not necessarily be due to the removal of water. However, the usual methods employed for the measurement and control of moisture content of cotton textiles described earlier are equally applicable to synthetics, though the range of moisture contents is usually lower except in the case of viscose rayon, which is regenerated cellulose.

4 MOISTURE SORPTION BY WOOL AND SIMILAR FIBERS

Wool has a higher moisture regain than any other fiber, being approached in this only by viscose rayon. The regain of wool in different forms has been measured in atmospheres of relative humidities ranging from 44 to 90% at a temperature of 72°F by Shorter and Hall [62], Speakman [63], and Goodings [64]. Different qualities of wool (and mohair) were investigated by them. It was found that the quality of the material had very little or no effect on the regain. One of the qualities (64s Cape wool) showed perhaps a significant divergence from the other four qualities. Thus at 90% relative humidity the regain of the Cape top was 21.30%, while those of the other four types were 22.95, 22.14, 21.90, and 22.00% respectively. It would not be safe, however, to generalize from this about the regain of Cape wools. Speakman [63] obtained the saturation regain values of wool as lying between 32.9 and 35.3% whereas Goodings [64] showed evidence of a higher saturation regain for his sample of wool, that is, 37.5%. Ashpole [65] carried out sorption experiments at relative humidities above 98% and showed that in this range the absorptive capacity of textiles in general and wool in particular is very high, the saturation regain was found to be close to the moisture imbibition obtained by standard centrifuging techniques. Oil was found to diminish the regain of wool. The oil apparently exerts practically no effect on the moisture-absorbing power of the wool. It produces a diminution of the calculated regain merely by loading the material with a nonhygroscopic substance.

It may be stated generally that the average moisture regain in various types of wool and woolen materials in an ordinary living room is as follows:

1. 15.5% for scoured wool
2. 15.5% for tops or yarn in oil
3. 16.0% for dry tops or yarn
4. 15.0% for noils
5. 15.0 to 17% for cloths

The previous history of wool fibers seems to have pronounced effect on the sorption characteristics and imbibition of moisture. It has been observed experimentally by Watt and Kennet [66] that the equilibrium weight of a wool sample obtained on desorption in vacuo is dependent on the previous treatment of the sample with respect to water. Exposure to 5% relative humidity has no effect on the subsequent equilibrium weight in vacuo, but exposure to intermediate relative humidities has appreciable effect. The maximum rate of increase occurs on exposure to approximately 50% relative humidity. Exposure to high relative humidities (near 100%) does not cause an increase in the subsequent equilibrium weight under vacuum. Initially the increase of equilibrium weight under vacuum is linearly dependent on the time of exposure to a particular relative humidity. It has also been found that increases are less marked at higher temperatures (65°C). Increase of the equilibrium weight in vacuo, however, can be removed by saturating with liquid water or water vapor prior to desorption. This phenomenon has been explained by Feughelman and Watt [67] by assuming that water molecules chemically bonded in the keratin structure tend to associate with the wool fiber by hydrogen bonding. This view is supported by the fact that experiments on the keratin-water structure, using nuclear magnetic resonance techniques, indicate an increase of mobility of the proton, and hence of the water molecules with an increase of relative humidity.

Water in excess of that present when a fiber is dried to zero percent relative humidity after equilibrium in water (100% relative humidity) is incorporated in the keratin structure when the fiber is dried after equilibration at lower relative humidities. The formation of this incorporated water is shown to be at a maximum

4 MOISTURE IN WOOL   613

after drying from about 50% relative humidity. From the discussion
of the state of water in the keratin-water structure, it appears
that at 50% relative humidity, we have the optimum situation of water
being mainly associated with the keratin structure and at the same
time there is sufficient mobility of the water and the structure to
allow the formation of optimum bonding of the water molecules into
the structure. When the fiber is equilibrated at 100% relative humidity most of the water present has only minor association with the
keratin structure. On drying far less rearrangement of the keratin-water structure can occur. Figure 9.9 shows the absorption and desorption curves of wool at various relative humidities as determined
by Speakman and Cooper [68].

It has been observed by Speakman and Cooper [68] that the affinity of wool for water decreases considerably with increasing

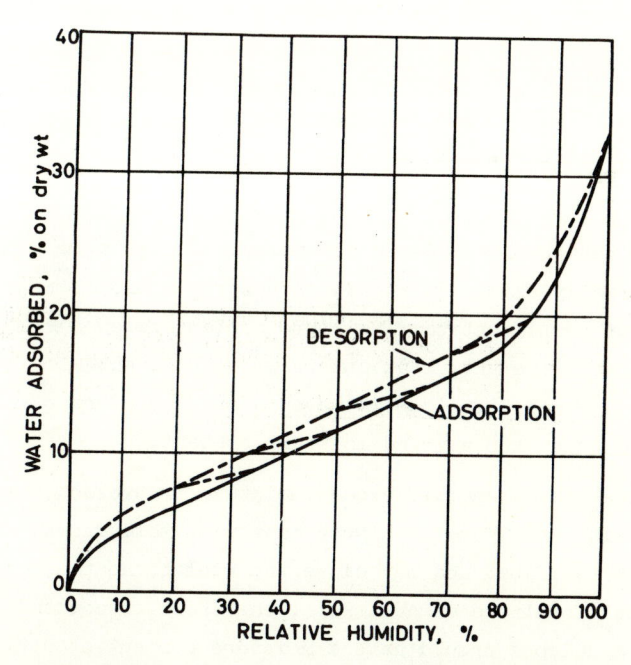

FIG. 9.9. Absorption and desorption curves of wool at various
relative humidities.

temperature at a given relative humidity. Hence for determining the equilibrium regain at a particular humidity, it is necessary to take the temperature factor into account. Similarly pH of wool also affects the equilibrium regain which becomes slightly higher in the alkaline state and lower in the acid state [68]. The pH effect is reversible provided wool has not been exposed to sufficiently extreme conditions to alter it permanently.

According to the British Standards specification [126] the "Standards of Official Regain" which have been proposed and are commonly, but not universally, accepted for the calculation of correct invoice weights from dry weights are as follows:

1. Scoured wool, 16% regain
2. Carbonized wool, 17% regain
3. Tops, dry combed, 18% regain
4. Tops, oil combed, 19% regain
5. Yarn (woolen), $18\tfrac{1}{4}$% regain

Regain is defined as

$$\frac{\text{Weight of moisture} \times 100}{\text{Weight of dry fiber}}$$

as determined in the Bradford conditioning house oven. Continental buyers normally allow 17% regain on scoured wool. Some purchasers will accept certificates of moisture content below 16% for calculation of invoice weights, but others prefer the actual weight to be up to invoice weight. Loss in revenue may therefore result if wool is shipped below the allowed moisture content.

On the other hand, if excess moisture is present, the purchaser may complain, and there is also a danger of mold growth which deteriorates the fiber and may cause the wool to heat. The Overseas Shipowners' Committee of Australia, therefore, recommends that clean wool being shipped should have a moisture content of not above 17% regain to avoid any risk of self-heating in the bale. Greasy wool should have this value reduced by a factor calculated from the grease content of the wool. For example, 20%-regain scoured wool usually has a grease content of approximately 1% so that 16.8% regain is regarded as safe. Slip wools generally have a higher grease content.

It is therefore important to the wool industry to control the moisture content of baled wool within fairly fine limits. Loss in revenue may occur if wool is shipped with less than optimum moisture content. On the other hand, if the moisture contents are higher than normal regain value, the purchaser may question the weight of bales, and there is also danger of staining or mildew damage, and of heat developing in the bales. Wet wool must, therefore, either be overdried and spread on a bin room floor, allowing it to pick up the proper amount of moisture in the course of a few days, or the dryers must be adjusted to give a wool of the correct moisture content which can be baled as soon as it has cooled sufficiently. For proper control, rapid and accurate estimation of moisture of wool is required.

4.1 Effect of Chemical Modifications

The ability of wool keratin to absorb water vapor depends on its chemical composition and molecular structure. The chemical reactivity of wool is largely dependent on the behavior of the disulfide bonds. The marked effect of the modification of the disulfide bonds on the mechanical properties of wool fibers has been clearly demonstrated by Harris [69]. Changes of the cystine content of wool by reduction or conversion into other products may therefore be expected to have some effect on the subsequent sorption behavior. Mellon et al. [70] have reported, however, that the reduction of disulfide bonds to fulfhydryl groups in wool does not affect the wool water-sorption isotherm. However, information is lacking on the kinetic behavior of water sorption by cystine-modified wools.

Watt and Leeder [47] have measured the water-vapor absorption of wool keratin after various chemical modifications in three humidity regions (low, intermediate, and high, i.e., less than 5%, 50 to 80%, and more than 80%, respectively) and have found that the mechanism of sorption varies with the initial water content of the wool. At low humidities (less than 5%), polar side chains have an important role, at intermediate humidities (50 to 80%) structural rearrangement of the fibers is significant, whereas at high humidities (greater than 80%) the observed kinetics depends on the ease with which conformational changes can occur. They concluded that specific sites should affect the lower region of the isotherm and that conformational changes should have a greater effect in the solution region. However, the modification of hydrophilic sorption sites by deamination and acetylation has no effect on the absorption isotherm, while conflicting results have been given for the water uptake of wool following methylation of carboxyl groups [71]. Decreases of water content have been reported following side-chain reactions, as well as reactions with isocyanates, and the addition of various acids. A reduction of swelling in liquid water was observed for wool fibers reacted with ninhydrin or containing dyes. Mellon and co-workers [70] have reported that no change in the isotherm of cystine-reduced wool takes place. Increased swelling, however, has been reported by Carter et al. [72] involving chemical modifications which disrupt keratin structure, such modifications leading to changes in water sorption.

The sorption behavior of wool fibers with reduced cystine content has been studied by Watt [73] using water and formic acid as absorbates. Reduction alone causes little change of the sorption behavior but reduction followed by alkylation of the sulfhydryl groups causes changes which depend on the extent of the reduction and the alkylation procedure. The water absorption isotherm of the alkylated wool is less than the corresponding isotherm of unmodified wool near saturation humidities, where the water content becomes higher for the alkylated samples. For absorption at high humidities the uptake curves of reduced and methylated samples show large overshoots of water content above the equilibrium water content value.

For samples alkylated with alkyd dihalides such that new crosslinks are formed, the rate of uptake is less than for unmodified wool, but at saturation the water content is between that of unmodified wool and reduced and methylated wool. The conversion of cystine to lanthionine in wool also leads to increased water contents at high humidities. The changes in sorption behavior of the modified wools are more marked when formic acid vapor is the sorbate. The effect of additional crosslinks on sorption by wool keratin has also been investigated by Watt [74], who has observed that the formation of additional crosslinks generally reduces the sorption of the vapors of water and formic acid, whereas the removal of the crosslinks between the peptide chains of keratin leads to an increase in the absorption of these vapors. Similar increases in the uptake of water vapor by keratin occur after peptide bond cleavage. When wool (keratin) fibers sorb water, the overall diametral swelling from dry to wet (zero to 100% relative humidity) is 16% while the crystalline regions swell by only about 5%, as judged by the 9.8-Å equatorial spacing.

4.2 Moisture Sorption by Wool Beds and Woolen Bales

When a stream of moist air is forced through a uniform bed of hygroscopic fibers such as wool, water is transferred from the air to the fibers or vice versa, depending on the relative humidity of the air and the concentration of water already present in the fibers (regain). Simultaneous heat transfer occurs causing temperature changes in the air and the fiber bed. It was shown by Cassie [75] that such a process is characterized by the propagation through the bed of two temperature fronts, one fast and the other slow. Due to his assumptions of infinitely fast heat and mass transfer rates, Cassie's theory predicted that the shapes of these fronts (with respect to time) were the same in the fiber bed as at the boundary. Later work by MacMohan and Downes [76] elaborated the theory to take into account a finite rate and mass transfer, the validity of which had been established since the publication of Cassie's original work. In solving the differential equations pertaining to the problem, MacMohan and Downes [76] assumed solutions in the form of fronts traveling through the bed at constant velocity. These assumptions necessarily limited the range of the solutions to cases where stable fronts are formed and to times when the steady state has been established. Nordon's [77] recent work on heat transfer in beds of nonhygroscopic fibers has led to a reexamination of the more general problem of combined heat and mass transfer in a wool bed.

The temperature rise produced by absorption results in an increase in the equilibrium pressure of the wool that is larger than the drop in equilibrium pressure due to the fall in temperature. Since the rate of mass transfer is proportional to the difference between the equilibrium pressure of the wool and the vapor pressure in the surrounding air, the subsequent rate of mass transfer is smaller for absorption than for desorption. The effects due to small changes are cumulative, resulting eventually in a completely different overall mode of transfer. These differences can also be demonstrated by examining the appearance of the temperature fronts for the two cases which are shown in Figs. 9.10 and 9.11. The dotted lines represent the approximate courses of the first temperature fronts, which are not covered by the Nordon model. The full lines represent the second fronts, which are associated with changes in concentration. These are the changes in temperature with respect to time as seen by thermocouples placed in the wool bed at various depths, and constitute a convenient basis by which the theory can be tested experimentally.

When the humidity or temperature of the atmosphere around a bale of wool changes to new levels, the regain in the superficial layers of the bale changes appreciably within a few hours, but at increasing distance into the bale, the change in regain becomes progressively less. Some weeks of exposure to the new conditions are necessary before there is much change in the regain of the wool near the center of the bale, and some months of exposure are necessary

4 MOISTURE IN WOOL

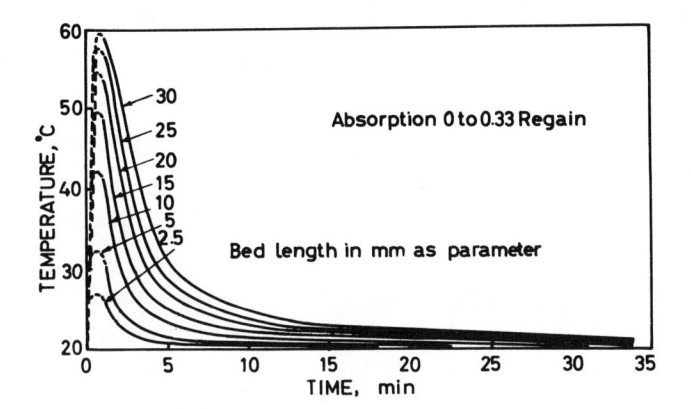

FIG. 9.10. Temperature-time relationships for various bed positions in a compressed wool bale in the case of desorption.

if the regain throughout the bale is to be substantially in equilibrium with the new atmospheric conditions. The times required depend on the dimensions and density of the bale. The slowness of the diffusion and the prolonged existence of the accompanying regain gradients are important commercial considerations in many circumstances. The gradients can lead to appreciable errors in conditioning certificates if the samples are not drawn from the correct portions of the bale.

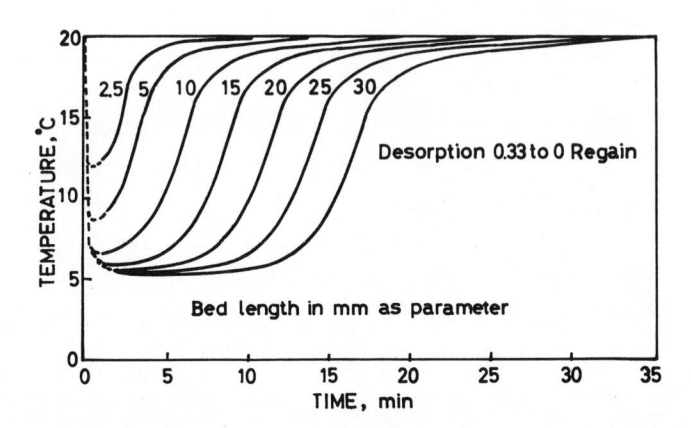

FIG. 9.11. Temperature-time relationships for various bed positions in a compressed wool bale in the case of absorption.

The theory of diffusion into absorbing media was developed some years ago by Henry [78] and a considerable number of observations have been made by the Shirley Institute on bales of cotton and certain fibers other than wool. Edmunds [79] has published some experimental results on bales of scoured wool in which gradients had developed while awaiting shipment, the outer portions of the bale acquiring a higher regain than the inner portions. Edmunds [79] observed that the difference between the surface and interior regains in such bales usually varied approximately in a simple cosine-law manner with the distance of the interior point from the center of the bale. He concluded that with such gradients there was, for any particular coring tube, a position in the end face of the bale through which cores taken would have almost the same regain as the whole bale. Edmunds [79] measured the regain gradients in the bale as existing (the prior regain history of the bale was not exactly known); consequently, he did not report the form of the gradient at various stages from its inception till its final disappearance.

Edmunds [79] has reported moisture gradients of 3 to 4% regain in commercial bales of scoured wool. His measured gradients confirmed the theoretical and experimental results obtained by Walker [80] who has conducted experiments in which changes in ambient temperature up to 49°C were suddenly applied to bales of raw wool, over a range of temperature. It has been shown by Walker [80] that the subsequent central temperature changes in the wool are in general accord with Henry's theory [78] for the coupled diffusion of heat and moisture. Walker has shown theoretically that a form of Henry's equation can be deduced to predict weight changes in baled wool when the ambient atmosphere has been changed, without requiring empirical determination of regain under new conditions. It has been found by Walker [80] that Henry's theory [78] is successful in some cases for predicting weight changes in wool bales suddenly exposed to changed atmospheric conditions. Nordon et al. [81] have studied the moisture changes in bales of greasy wool and have suggested that it is better to test the relative humidity inside the bales rather than the moisture content.

4.3 Moisture Regain and Dimensional Stability of Wool Fabrics

The relation between dimensional stability of wool fabrics and regain has been investigated by a number of research workers. The dimensions of wool fabrics nearly always vary with the regain in a reversible way, whether or not any type of shrinkage takes place at the same time. This is a well-established fact, it being generally true that an increase in regain causes an increase in both length and width, corresponding contractions taking place when the regain is lowered. However, reversible change of dimensions in wool fabrics has been reported by Von Bergen and Clutz [82]. The reversible effect has been called RH motion by Von Bergen and Clutz [82], reversible dimensional change by Cednas [83], and hygral expansion by Baird [84-85].

Bekku [86] has reported experiments in which wool fabrics exhibited hygral expansion, Nutting [87] has dealt with the phenomenon as observed in a knitted structure, and Baird [84-85] has described how it affects some aspects of relaxation shrinkage. The cause of hygral expansion in woven fabrics has been explained and investigated by Cednas [83] and Olofsson [88-89] whose work has clarified this behavior considerably. The hygral expansion of a number of worsted fabrics has been studied by Baird [84-85], who has shown that the yarn and fabric behavior can be explained in terms of the bending and straightening of single fiber arcs. The hygral expansion curves for three worsted fabrics set in water at different temperature are shown in Fig. 9.12.

As regards a theoretical explanation of the hygral expansion of wool, there is some controversy between views held by different investigators. The reader interested in this theoretical interpretation is referred to the original publications [88-89] on the subject, as it is beyond the scope of this book. The problem of structural changes accompanying the setting and supercontracting of modified wool fibers along with the phenomenon of hygral expansion of wool fabrics has been studied by Haly and Snaith [90]. The two-phase theory for wool advanced by Feughelman [91] together with the

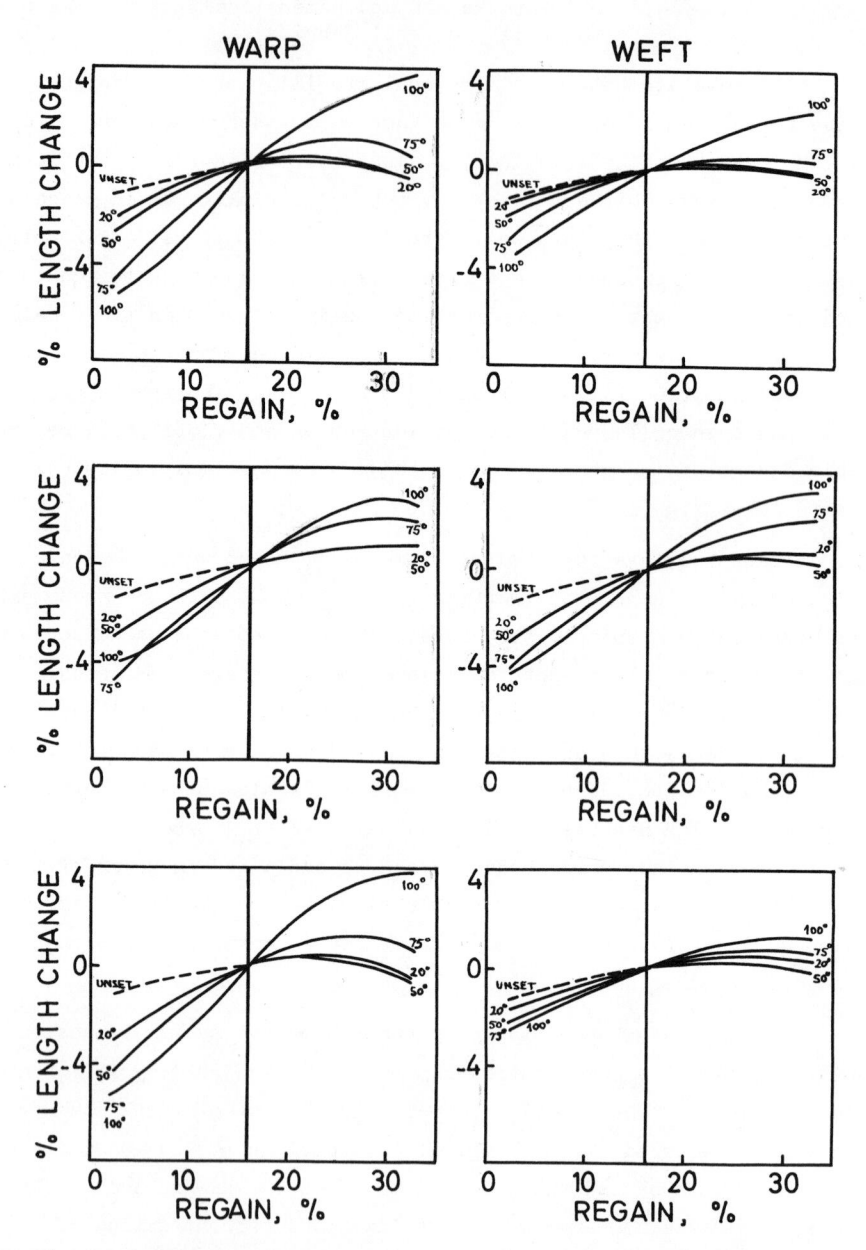

FIG. 9.12. Hygral expansion curves for three worsted fabrics at different temperatures.

model for changes during setting proposed by Feughelman et al. [92] appears to be adequate to explain the results obtained by Haly and Snaith [90].

4.4 Moisture Sorption by Specialty Hairs and Fur Fibers

The moisture regain of wool as a function of relative humidity, under both adsorption and desorption conditions, has been well established, but very little information is available concerning the moisture regain of keratin fibers other than wool. In manufacturing circles the assumption is usually made that the less common keratin fibers (such as specialty hair and fur) closely resemble wool in their moisture regain properties. Since the importance of the hygroscopic properties of wool is well established, one would expect equal or greater importance to hold for the costlier fibers such as cashmere, vicuna, and Angora rabbit fur. Moisture regain data for a number of the less common keratin fibers have been reported by Von Bergen [93]. These can be summarized as follows:

1. The moisture adsorptive and desorptive powers of the specialty hair fibers are remarkably similar to wool. The same hysteresis exists in the moisture content of mohair, cashmere, alpaca, vicuna, and camel's hair fibers between adsorptive and desorptive conditions.

2. The moisture regain of fur fibers such as rabbit, beaver, and muskrat is approximately 2% lower than wool and the specialty hair fibers.

3. At 65% relative humidity and 70°F the average moisture regain of the specialty hair fibers is about 15% and that for the fur fibers is 13%.

4. Since the fur fibers are finer (about 14 μm in diameter), the affinity for water is decreased. This fact is in agreement with Speakman's [63] observation that affinity of wool for water appears to increase slightly as the wool is coarser.

5. The moisture regain of fur fibers does not seem to be much affected by the various chemical treatments such as pulling, boiling in acid, and bleaching.

4.5 Theory of Moisture Sorption by Wool

Most of the discussions of the moisture regain versus relative humidity isotherm of wool and analogous materials have been based on the principle advanced by Pierce [94] that at the lower relative humidities water is firmly bound to specific sorption sites, while at the higher relative humidities the adsorbed water is loosely bound. An excellent discussion of developments up to 1946 is found in the papers of Cassie [95] and Gilbert [96]. At the present time the theories of Katz [97], Pierce [94], Brunauer et al. [98], Cassie [99], and Hill [100] are in vogue. The isotherm is derived by consideration of an evaporation condensation mechanism; however, the statistical thermodynamic theories of Cassie [99] and Hill [100] yield the same isotherm. The approach due originally to Katz [97] involves initial hydrate formation (absorption on sites) followed at higher relative humidities by an approximation to a solution mechanism, characterized by an increase of entropy with increase of water content.

McLaren and Rowen [101] have reviewed some further developments, in which solution theory has played a prominent role. Hailwood and Horrobin [102] assumed that an ideal solid solution was formed in the wool-water system, while Rowen and Simha [103] adopted the solution theory of Flory [104] and Huggins [105]; they introduced into their equation for the free energy of water, a term arising from the elastic constraint of the polymer network, as had been done by Cassie [99] in deriving the stress-free isotherm.

The theoretical aspects of sorption and desorption of moisture by wool have recently been reviewed thoroughly by Sule [106] in which the divergent opinions advanced by different researchers have been discussed. There is at present no general agreement as to the formal theoretical approach. While an analysis may be successful in many respects, it fails in others. For example, Cassie's treatment led to calculations of quantities of bound and free water which correlate with some experimental data, though this subdivision was found inadequate by Windle and Shaw [107]. None of the treatments provide a sufficiently detailed model of the range of binding energies involved to account for the more recent experimental finding that the quantity of incorporated water in wool depends on previous sorption history. A recent empirical analysis by Feughelman and Haly [108], while greatly simplified in some respects, does give an explanation for the phenomenon.

With respect to site absorption there has been a number of investigations to show that change of the polar environment in wool or other proteins may decrease absorption. Mellon et al. [109] found substantial decreases following benzoylation of casein, as did Moore [110] following treatment of wool with isocyanates. On the other hand, Speakman [63] obtained an unaltered isotherm for deaminated wool. Nicholls and Speakman [71] observed a small decrease due to methylation, and in some cases quite large decreases following the treatment of wool with acids, these decreases are only partially due to changes in site absorption. Haly and Snaith [111] have published swelling data which show that water contents as high as 70% are reached by fibers which have been supercontracted in lithium bromide solution and washed. The isotherm for wool treated in this way begins to deviate appreciably from the normal isotherm only above the relative humidity of 85%.

All macromolecular systems give isotherms which represent combinations of two of the following three processes.

1. Solution sorption as per Henry's law [78]
2. Van der Waals' adsorption as per the Langmuir isotherm
3. Multimolecular adsorption

The multimolecular theory, known as the BET theory, is now generally accepted, as it is able to explain the characteristics of different types of sorption isotherms. According to this theory a monolayer is considered to hold strongly on specific sites which provide secondary sites for the condensation of more molecules in the form of multilayers. Katz [97] has advanced a solution theory for explaining sorption of water vapor by polymers. A number of workers criticize the application of absorption isotherms to polymer-water systems. This criticism is based on the ground that such theory can be strongly applied to systems with a definite rigid surface and that it is meaningless to talk of the internal surface of a polymer. Therefore, an alternative approach had to be adopted based on the solution of polymer in the solvent. In this respect Hermans [112] has supported the interpretation of Katz [97] that after the initial hydrate formation, the water absorbed at higher regain forms a solid solution [113-114].

The relationship of water content to relative humidity has been discussed briefly by Haly [115]. He observed increased saturation of water content for wool fibers which were supercontracted, reduced, and sulfur methylated. According to his observations (in the case of sulfur-methylated wool) saturation water contents show little or no dependence on the number of sulfur methyl groups present. His results have been interpreted in terms of the Flory [104] and Huggins [105] solution theory, and it is indicated that this theory can be applied usefully for relative humidities in excess of 85%. In normal wool solution absorption is permitted only in the noncrystalline component, but in supercontracted wool or sulfur methyl wool it takes place in both components. Even when saturation water content is greatly increased, the isotherm is not appreciably changed below 85% relative humidity, indicating that no additional sites for strong binding of water become available; apparently all such sites are accessible in normal wool at approximately 85% relative humidity.

In a brief communication, Morrison [116] reports that in the regain of 35 to 42%, water is thermodynamically bound to wool keratin. His proposal is based on the results of experiments in which water vapor is slowly withdrawn via a capillary tube from a flask holding a wool sample initially containing free water, air being absent. Downes and Nordon [117] are of the view that water is not bound to keratin in the proposed range and that Morrison's observations result from a failure to maintain the temperature throughout his sample sufficiently uniform. Their conclusions are based on theoretical grounds and are supported by results they have obtained from an experimental arrangement similar to that described by Morrison [116]; however, they used a much reduced evacuation rate to obtain more nearly isothermal conditions.

Watt [118] has proposed a two-stage absorption mechanism for wool-water systems. He found that the kinetics of the uptake of water vapor by wool is dependent on the initial concentration of water and the size of the concentration increment and initial regain of the wool. For small concentration increments, absorption may occur in two stages, the second stage being much slower than the first. It has been shown by him that first-stage absorption to a quasiequilibrium state obeys Fick's law of diffusion with a concentration diffusion coefficient, and the activation energy varies with concentration. Second-stage absorption is accompanied by irreversible configuration changes within the fibers. As an example it was observed that for low regains (zero to 2.5%) sorption is Fickian with a concentration-dependent diffusion coefficient. At 35°C and zero regain the diffusion coefficient is 11 kcal/mole. For concentrations above 2.5% regain, a portion of the water-vapor uptake on absorption occurs by a non-Fickian mechanism; the relative contributions of both the Fickian and non-Fickian mechanisms to the total uptake vary with the initial regain. The correlation between mechanical properties of wool such as stress-strain characteristics and regain values have been investigated [119-120] at the CSIRO Wool Research Laboratories (Australia).

4.6 Methods of Moisture Measurement of Wool

The five principal methods for determining moisture in wool are as follows:

1. Oven-drying or gravimetric method

2. Chemical methods, including the Karl Fischer method

3. Measurement of the humidity of air in equilibrium with wool

4. Measurement of the changes in electrical properties of the moist wool

5. Measurement of the nuclear magnetic resonance property of protons of wool moisture

Before describing the oven-drying method, it is important to point out that it is rather difficult to find the exact bone-dry weight of wool. The difficulty is perhaps greater in the case of wool than in the case of other fibrous materials. While determining the vacuum dry weight of a sample of wool, it was observed by Watt and co-workers [121] that equilibrium weight of the sample is dependent on the conditions of desorption as well as on its previous history with respect to its contact with water vapor. Watt and Kennet [66] have shown that water may be incorporated into the keratin structure by long exposure at relative humidities of between 5 and 80%. This incorporated water is not removed on drying in vacuo and results in an increase of the dry weight of the wool fibers. Mechanical tests carried out on fibers containing incorporated water show an increased stiffness for relative humidities of less than 60-70%. It has been proposed by Feughelman and Watt [122] that the water which is incorporated into the keratin structure and increases the dry weight becomes immobilized at intermediate relative humidities by increased strength of bonding of the water directly with the keratin structure. Saturation in liquid water reduces the amount of association between one molecule and the keratin structure. Drying rapidly from this state, therefore, lowers the possibility of formation of strong, direct bonds between water and keratin. According to Watt and Kennet [66] the equilibrium weight of a wool sample obtained on desorption in vacuo is dependent on the previous treatment of the sample with respect to water.

Recently, drying units have been developed [123] which dry wool samples in about 7 min. By using such a drying equipment and a vibroscope (which enables instantaneous measurement of regain), Mackay [124] has determined the drying time of a wool fiber heated at 105°C. He found that the time for the single fiber to dry to 0.5% regain when exposed to dry air at 105°C was not more than 1.2 min and to reach 0.1% regain, not more than 3 min. By the International Wool Textile Organization [125] definition of dryness (namely, that the rate of change of weight is less than 0.05% per 15 min), the fiber took 4 min to dry.

The method of drying and weighing is based on the British Standards Specification [126] and estimates the water content of 500 g of wool (1 lb) in approximately 1 hr. Though this method is quite accurate and officially recognized in the industry for the purpose of making invoices, it takes rather too long a time, and therefore quicker methods of moisture measurement using oven-drying techniques have been developed. The CSIRO direct reading regain tester (Fig. 9.13) is one such equipment that has recently been developed especially for wool by Mackay [127] which can accurately estimate the moisture content of a 100-g sample of wool in about 5 min.

Figure 9.14 shows drying curves obtained for various samples of wool at room temperature, and, for comparison, the characteristic drying curve of a single wool fiber is also given. By the use of such drying curves, a time of drying can be selected appropriate to the accuracy desired; e.g., the 292-g sample of top was dried to within 0.25% of equilibrium in under 3 min and to 0.1% in 6 min and the 250-g sample of scoured merino to 0.1% in 3 min. Observational errors are unlikely to exceed 0.1% of regain.

It was found that when used on an oil-free sample, the CSIRO tester gives results which, with reasonable certainty, lie in the range of ±0.5% of the true regain of the sample. If a correction is made for the moisture content of the air, the limits become ±0.32% regain. However, recent tests carried out by Mackay et al. [128] have cast doubt on the validity of the earlier results

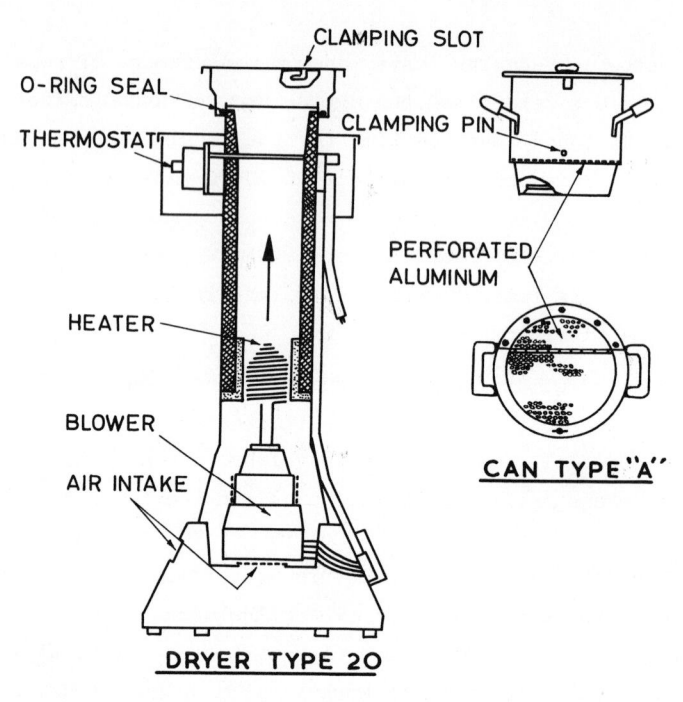

FIG. 9.13. Schematic diagram of the dryer and the air heating and mixing arrangement.

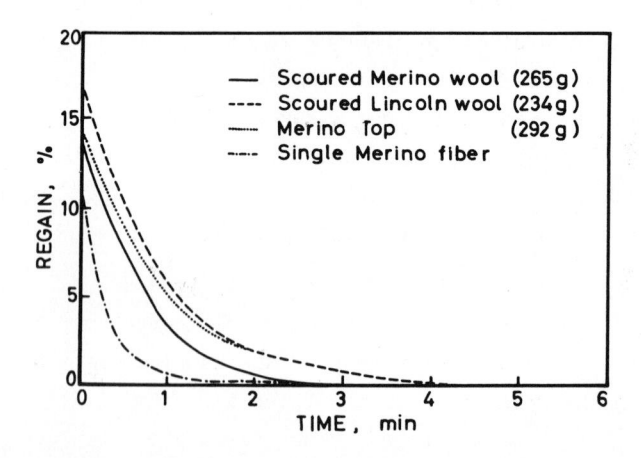

FIG. 9.14. Drying curves of various samples of wool.

obtained by them. A 3-lb sample moisture meter for the rapid control of moisture in wool from machine dryers has been developed by Rothbaum and Vera-Jones [123] of the Dominion Laboratory (New Zealand). This moisture meter is useful for wool scourers. It is one of the instruments used for precise control of moisture in the processing of wool. Figure 9.15 shows the general arrangement of the apparatus, though some modifications would be required for alternative types of heater or fan. It consists essentially of four parts: (1) a container for the wool, (2) a fan to provide the flow of drying air, (3) a heater to heat the air, and (4) calibrated scales for weighing regain directly. For detailed descriptions of these individual units, the reader is referred to the original publication [123].

Watt et al. [129] reported that the dry weight of wool samples determined in air at 105°C depends on the previous sorption history of the sample. In particular, samples which had been kept at room condition for some months before drying showed a dry-weight reduction of about 0.4% when wetted out and redried. Such a result, if confirmed, would be of considerable importance in commercial determinations of dry weight. Its magnitude is very significant in relation to the tolerances usually allowed in determining regain for certificate purposes.

LeCompte and Lipp [130] have used vacuum drying over phosphorus pentoxide as a comparative method for determining moisture in wool. They found that vacuum drying over phosphorus pentoxide can give complete or total moisture content provided sufficient time could be allowed for the moisture-laden wool to give up all its moisture and thus come in equilibrium with unreacted phosphorus pentoxide, which theoretically allows no moisture to remain in the atmosphere around it. However, it is very difficult to attain this condition in practice. They observed that the samples were losing moisture at the rate of 0.05% every 24 hr and that constant weight was approached in at least 62 hr. The values obtained by them were rather low. Although theoretically vacuum drying over phosphorus pentoxide

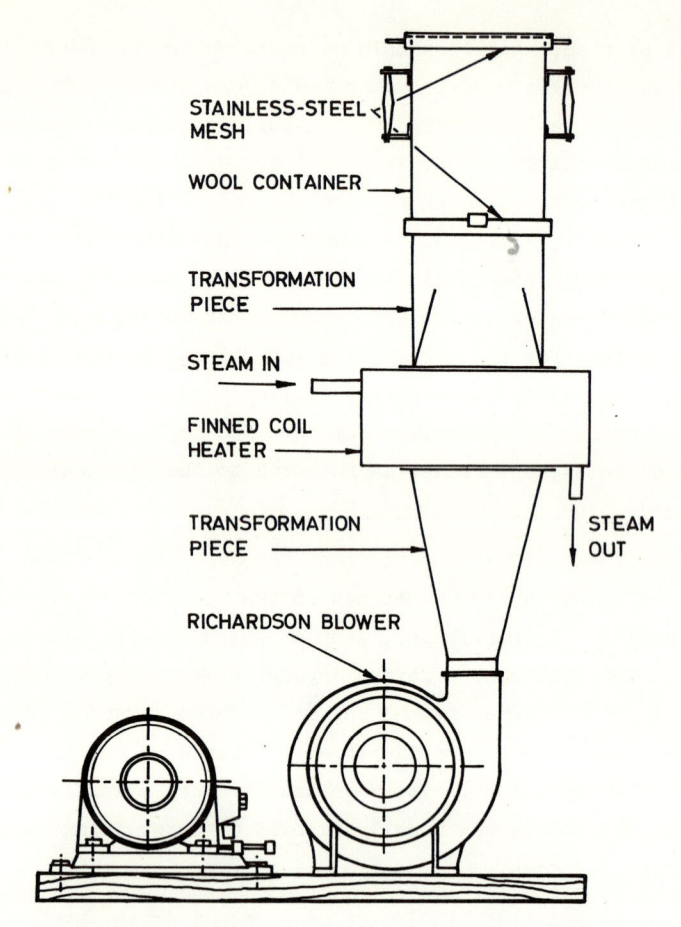

FIG. 9.15. Schematic diagram of the Dominion Laboratory (New Zealand) meter for wool.

is a faultless method, practical applications offer considerable difficulties. They came to the conclusion that accurate and convenient testing for moisture in wool would depend on the development of a conditioning oven which would operate at elevated temperatures and at the same time be supplied with a continuous supply of anhydrous air.

Chemical methods generally useful for the measurement of moisture content of other hygroscopic materials are equally suitable for wool. The Dean and Stark [131] azeotropic distillation method and the Karl Fischer titration techniques [132] have been applied with success for the measurement of moisture regain of wool. LeCompte and Lipp [130] have used toluene distillation techniques for the measurement of moisture content of scoured wool. They observed that toluene distillation method gives very accurate results for wool.

Wilson and Sandmoire [133] have recently used the Karl Fischer reagent for the determination of the moisture content of woolen fabrics as well as of nonwoven wool. They have also compared these results with the ASTM conventional oven method [134] and vacuum oven methods and have analyzed the results statistically. These researchers have replaced methanol by formamide as the water-extracting solvent which according to them offers the advantages of precision, accuracy, and speed. The volume of formamide used must be large enough to cover the sample completely. As is well known it is a powerful dehydrating agent, capable of dehydrating activated alumina, silica gel, and such other desiccants. It was found by these investigators that on direct titration formamide gives a more stable end point than methanol. The moisture contents of the following woolen materials were determined by them [133]:

1. Wool-nylon flannel (20% nylon) dyed olive green
2. Yellow wool flannel
3. Wool flannel, bleached (H_2O_2)
4. Wool flannel, severely bleached (H_2O_2)
5. Wool flannel, unbleached
6. Wool flannel, 1% modified epoxy resin
7. Wool flannel, 3% modified epoxy resin
8. Wool having grease
9. Mohair having grease
10. Top wool
11. Card sliver wool
12. Mohair, scoured
13. Clean loose wool

Table 9.5 shows the average moisture values experimentally determined. Each average represents triplicate determinations on two

TABLE 9.5

Average Moisture in Cloth and Nonwoven Wool Obtained by Different Moisture Methods and Methods of Preparation

Material	ASTM procedure Prepn. A	ASTM procedure Prepn. B	Vacuum oven method Prepn. A	Vacuum oven method Prepn. B	Karl Fischer method Prepn. A	Karl Fischer method Prepn. B
Woolen fabric						
1	8.51	9.01	8.60	9.07	8.87	9.24
2	10.52	11.03	10.47	10.91	10.71	11.30
3	11.42	11.37	11.15	11.17	11.61	11.60
4	11.98	11.57	11.72	11.32	12.12	11.87
5	12.05	11.60	11.88	11.42	12.23	11.86
Nonwoven wool						
6	-	12.88	-	13.14	-	13.14
7	-	10.47	-	10.67	-	10.67
8	-	11.30	-	11.33	-	11.60
9	-	11.45	-	11.46	-	11.69
10	-	11.34	-	11.30	-	11.53

portions of material. Statistical analysis of these data for each of the three methods was carried out to study the relative accuracy of the methods and to determine the relative number of portions needed to provide a given level of precision for the average moisture content of material pieces. A standard deviation was computed for each set of observations and tabulated according to the moisture determination method and the method of preparation of the sample. It was found that a precision of ±0.1% is obtained with two portions, whereas seven portions give a precision of ±0.05%. Increasing the size of the portion in the Fischer determination also increases the precision. The volume of formamide used will also have to be increased in proportion to the increase in sample size. The volume of formamide used must be large enough to cover the sample completely. According to these research workers the Karl Fischer method of moisture determination for wool has several advantages over oven-drying methods. The method is much more rapid, as it takes only 10 to 15 min/sample for moisture determination; also it has higher accuracy, as the value of moisture content determined represents water present in the sample rather than the total loss in weight. This method should prove valuable as a calibration method for the determination of moisture in different types of wool and woolen clothing and fabrics.

Titration methods based on the reaction of water with acetic anhydride appear most promising for the determination of moisture content of cellulosic materials and wool. This method was originally developed by Mitra and Venkataraman [135]. A Speedy Moisture Meter (depending on the reaction of water in wool with calcium carbide) is commercially available, and appears reliable for small samples of wool.

Another method of measuring the moisture content of wool depends on the determination of ambient relative humidity in equilibrium with the wool sample. The measurement of the humidity of the air in contact with the wool is easily performed, and has the advantage that it could be continuously carried out and used to activate an automatic controller for the dryer. It is also scientifically of great interest, as it is the humidity which determines microbiological attack. However, the relations between humidity and moisture content vary somewhat in different wool samples, depending on grease content, on whether the wool was previously dry or wet (hysteresis), on the extent of drying it has undergone, and also on the time elapsed since drying. If the chief aim of the method is to give moisture content of the wool, it may give unreliable results. A commercial instrument depending on this principle is available and an improved version (the Telemeter) has recently been developed. This instrument gives rapid readings on samples of wool of any weight, but the values are strongly affected by the kind of wool used and the length of time elapsed since leaving the dryer. It also needs some skill on the part of the operative to control the wet and dry thermocouples. The details of instruments developed for the measurement and control of humidity of air are given in standard books on hygrometry, and the reader is referred to one such recent book by the author [136].

The electric and electronic moisture meters have been found to be quite satisfactory for the measurement of the moisture contents of cellulosic and synthetic materials. However, they do not seem to be very useful and accurate for the measurement of the moisture regains of different kinds of wool, especially in the case of greasy wool, as they give rather erratic results with different types of wool, depending on its fat content and other impurities. Lemann [137] has shown that between 10 and 40% regain the electrical resistance of pure wool is a good measure of its moisture content. Few instruments depending on this principle have been developed as yet, and it seems probable that residual slipping chemicals, grease, and other impurities would greatly affect the measurement. There are only a few publications describing successful methods of determining the moisture content of greasy wool by measuring its electrical parameters. Roberts and Algie [138] have measured the electrical properties of eleven wool samples of varying compositions over a range of humidities in order to find the magnitude of errors likely to occur when deductions of regain from electrical parameters are attempted. The errors involved in attempting to use the variation with regain of the dielectric constant or of the dc or ac conductivity of greasy wool as a measure of its regain have also been investigated by these researchers. It is concluded that in the frequency range up to 7.8 MHz these parameters are so dependent on the nonwool constituents of the greasy wool, particularly its suint content, that they are not likely to be useful for the measurement of the regain of wool. These investigators also used a wide range of frequency variation, i.e., from 150 kHz to 5 MHz, to find the effect of the change of the frequency on the dielectric constant of the wool at different values of regain conditioned in a wide range of relative humidities. Similarly the measurements of resistance of the samples of wool about 200 g in weight were made keeping the electrodes 1 in. apart. It was observed by them that suint has a larger effect on the variation of the dc resistance than on the dielectric constant (Table 9.6).

The dc resistance varies as much as 10^4 ohms while as stated earlier, a change of 2% in regain (e.g., 44 to 65% relative humidity for some samples) may cause a change of small fraction only. Suint consists largely of potassium salts of water-soluble organic acids, is deliquescent at 65% relative humidity and as such would be strongly ionized, so the specific conductivity of the suint together with its absorbed water at a certain humidity is expected to be higher than that of wool by a number of orders.

Although the proportion by volume occupied by suint in greasy wool is on the average about 10% of that of wool, and its spatial distribution is nonuniform, it has been found to produce a very large

TABLE 9.6

Dielectric Constant and dc Resistance Values of Greasy Wool Having Different Amounts of Suint Content

Sample	Suint (%)	Regain (%)	Effective dielectric constant (5 MHz)	dc Resistance (ohms)
B	0.3	14.8	1.79	1.5×10^{10}
C	2.1	15.1	1.85	4.5×10^8
D	3.2	11.5	1.99	6.2×10^8
E	3.2	12.2	2.02	2.6×10^9
F	5.8	13.1	2.26	7.5×10^7
G	6.2	13.9	3.98	1.9×10^5
H	7.5	13.1	2.45	9.2×10^5

(at 65% Relative humidity)

variation in the dc conductance of greasy wool and to have only slight effect on its dielectric properties. Orientation and distribution of the fibers between the electrodes affect the electrical parameters so we must expect the spatial distribution of the suint in the greasy wool to affect these parameters also, and, if the same suint content is distributed differently in different samples, or on repacking one sample, this can also be expected to produce differences in the electrical parameters. Roberts and Elgie [138] have also tested a commercial dielectric moisture meter for the measurement of the moisture content of greasy raw wool. This moisture meter works on the well-known principle of the variation of the effective dielectric constant of a capacitor in a circuit resonant at 7.5 MHz. Fairly accurate results were obtained with clean wool, but when the meter was tried on greasy wool there was a large variation observed between the meter readings and the moisture content as actually determined by the standard oven-drying method. In some cases the meter gave errors of over 100% in its indication of regain, as shown in Table 9.7.

In connection with the measurement of the regain of wool by dielectric-type moisture meters, Algie [139] has investigated in detail the effect of temperature on the dielectric constant of wool and the consequent corrections required in dielectric-type moisture meters. He has also studied the dielectric properties of wool top, and estimated the effect of impurities on the dielectric properties as a result of these two factors: (1) the variability of salt content between balls of top; (2) the effects of a nonuniform temperature different from that used in the calibration of the meter. The first of these causes an error of approximately ±1% at 15% regain (if the salt content is not known). The effect of a difference in temperature of ±2.5°C from the calibration temperature can cause an error of ±0.4% regain. The combined effect of all the sources of errors including the inaccuracy of the meter itself for use on all types of undyed top is approximately ±1.5% at 15% regain. However, a knowledge of the salt content would increase the accuracy to ±1.1%.

TABLE 9.7

Comparison of Regain Values by Commercial Dielectric Moisture Meter and Standard Regain Tester for Samples of Wool-containing Suint

Sample	Suint content (%)	True regain (%)	Indicated regain (%)
1	0.3	14.8	15.5
2	2.1	15.1	16.3
3	3.2	11.5	14.7
4	3.2	12.2	15.0
5	5.8	13.1	18.4
6	6.2	13.9	37.5
7	7.5	13.1	43.1
8	9.2	16.1	44.5
9	10.5	17.0	61.3
10	10.5	14.9	37.8
11	10.6	15.1	39.8

Application of a dielectric-type moisture meter (SRI Moisture meter) in conjunction with a special needle-type electrode has been made for the measurement of moisture content of highly compressed bales of raw wool and wool tops in a woolen mill by Pande [140]. The details of the electronic circuitry of this moisture meter have been described in Chap. V and, therefore, a short description of the electrode system only as applicable to the bales of wool and wool tops is given here. The needle-type electrode consists of three rows of tapered 5-in. needles arranged in such a manner that a double condenser is formed in parallel between the first and the second, and the second and the third rows of needles (Fig. 9.16). The central row of needles forms the common or negative point of the dual condenser. The positioning of the needles is arranged in such a preferred manner with respect to the handle of the electrode that a maximum thrust for piercing the compressed bale of raw wool is obtained as shown in the figure. This electrode system is con-

640   IX. MOISTURE IN TEXTILES

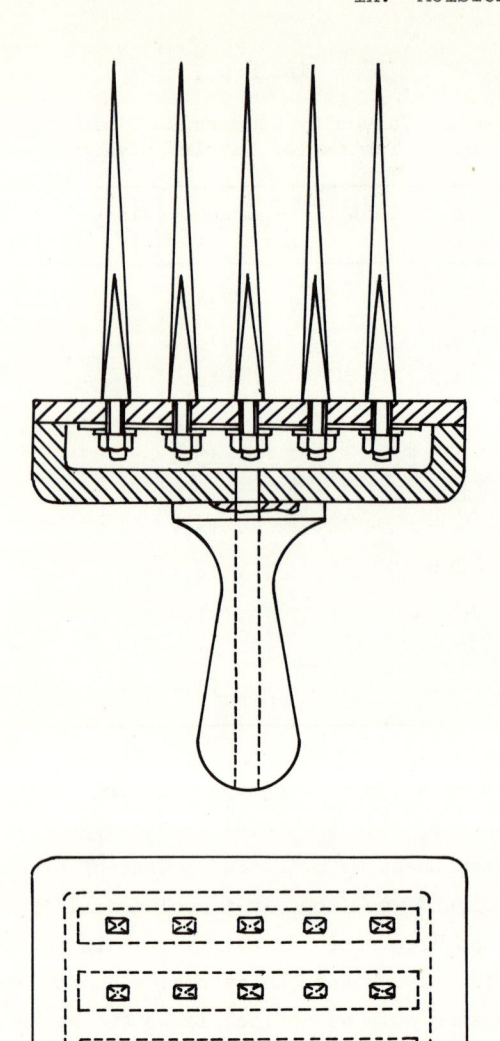

FIG. 9.16. Schematic diagram of needle-type electrode for wool bale.

nected to a coaxial cable which is in turn connected to the electronic unit of the moisture meter; the electrode along with this cable is completely matched to the electronic unit.

This electronic unit together with the electrode system has been calibrated on the spot in a woolen mill by taking core-bored samples from the compressed bales of wool at exactly the same depth and spot where a few moments earlier the needle electrode was pierced for the measurement of the moisture content by the electrical or electronic moisture meter. It was observed that a fairly close agreement, within 1% moisture content, between the moisture content values and the readings of the electronic moisture meter could be obtained if a calibration is made after obtaining a linear relationship between the moisture content of the raw wool, as obtained by the oven-conditioning methods, and the output current in milliamperes as indicated by the SRI electronic moisture meter. Such a moisture meter after proper calibration in situ has been found to serve the requirements of the industry for a rapid and accurate measurement of the moisture content of raw wool in a highly compressed form as in a bale. It was found, however, that grease, suint, and other impurities gave discordant results.

The effects of absorbed water on wool are so important and so general that in practically no branch of wool science is it permissible to neglect the state of hydration of the material. It has been shown that nuclear magnetic resonance techniques can also be applied for the determination of the nature and amount of water in wool. A publication by Shaw and Elsken [141] deals with some aspects of the wool-water complex. Both direct absorption curves and derivative curves due to proton magnetic resonance absorption in wool have been obtained by West et al. [142]. Peak heights and widths of the absorption lines were measured for wool samples equilibrated at a number of relative humidities in the range of 0 to 90%. All water protons in wool appear to be less firmly bound than the keratin protons, but less mobile than liquid water. It seems clear that an increment of absorbed water changes the binding energies of previously absorbed water. No evidence, however, was obtained of a subdivision of water in wool into two or three fractions with different binding energies which are successively bound as the equilibrium regain is increased.

Basically the method relies on the fact that the absorption line from protons of the absorbed water is narrower than that from protons of the wool keratin, and by suitable arrangement of the experimental conditions the lines can be effectively separated. For a detailed description of the experimental procedures, the reader is referred to Chap. VII.

The techniques used by West et al. [142] to obtain information about the resonance absorption fall into two general classes. In the first case the magnetic field is varied linearly with time over a range greater than the line width and the output from the detecting apparatus is proportional to the absorption. Figure 9.17 shows the shape of the line obtained from a material such as wool. The narrow central line is due to the absorbed water and this is superimposed on the very broad line due to the protein. The broad line is usually so small in amplitude that it becomes indistinguishable from the general noise in the apparatus and only the sharp, narrow line is obtained. In the second method the magnetic field is varied sinusoidally over a range which is small compared with the line width, while the frequency (or magnetic field) is varied linearly to sweep through the absorption band. The output is then a close approximation to the first derivative of the NMR absorption line. To obtain an adequate signal from the broad line the modulation must not be so large that it causes distortion of the narrow line. Hence it is necessary to use different values of modulation, depending on which line is being studied. When the modulation is reduced to a very low value (a small fraction of the width of the narrow line) no effective signal is obtained from the absorbent, so that the curve obtained is essentially that from the absorbed water.

Both of the methods of detection described above were used by West et al. [142] to obtain the absorption signal directly, a radio-frequency bridge circuit resonant at 30.5 MHz was used, and the absorption line was displayed on a cathode-ray oscilloscope. A perma-

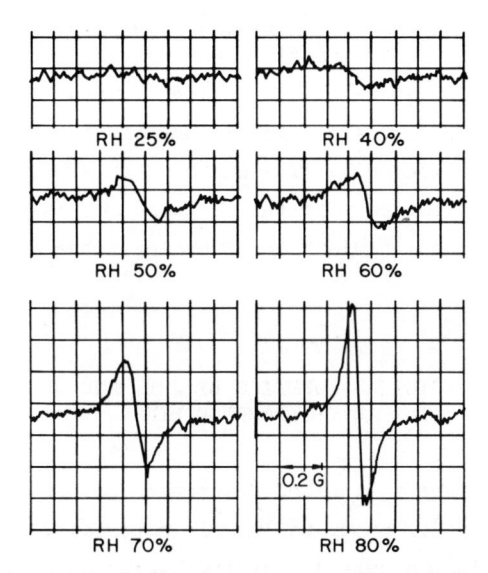

FIG. 9.17. Nuclear magnetic resonance lines obtained with a sample of wool.

nent record was obtained by photographing the oscilloscope screen. The magnetic field was increased linearly over a range of 0.85 G at a rate of 2 G/sec, and the radiofrequency field within the specimen was approximately 0.01 G to minimize the effects of saturation. For each line, the peak height and the linewidth at half the maximum amplitude were measured.

A Pound-Knight [143] spectrometer operating at 17.1 MHz was used to obtain the derivative curves. The magnetic field was modulated sinusoidally at a frequency of either 50 or 280 Hz. For most of the water lines in wool, the modulation amplitude was 20 mG peak to peak, but for some specimens with low water content the value was increased to 200 mG; this was still low relative to the line width. To obtain the protein line in the case of dry wool, the modulation field was 1 G; the radiofrequency field was approximately 0.02 G. For each derivative curve the peak-to-peak height was measured, as was the linewidth ΔH between peaks on the derivative curve. The specimen coils for both types of tests were approximately 7 mm inside diameter and 6 mm long. All the experiments were done by these investigators in a magnetic field provided by a Varian 12-in. electromagnet. To measure the variation of magnetic field throughout the specimen volume, a similar volume of distilled water was placed in the specimen coil. The width of this proton resonance line is due almost entirely to the variation of magnetic field. In the present case the linewidth ΔH at one-half the maximum height was 18 mG.

5 MOISTURE SORPTION BY JUTE AND SIMILAR VEGETABLE FIBERS

The moisture absorption and desorption properties of jute fibers are similar to that of other cellulosic materials. However, there are differences in the details, as the hygroscopicity of jute and other vegetable textile fibers is dependent on their chemical compositions. It has been observed that the moisture sorption characteristics of long vegetable textile fibers appear to depend to a great extent on the relative amounts of α-cellulose, lignin, and hemicellulose, the three principal chemical constituents of the fibers. Change in the moisture sorption of such a fiber due to a change in temperature of the surrounding air (its relative humidity remaining constant at 65%) is on the average 0.5% per 10°C.

Changes in moisture regain of jute after radiation with ^{60}Co γ rays up to 10^8 rad have been observed recently by Majumdar and Rapson [144]. An unusual but definite increase in moisture regain values was observed at 100 Mrad. It was found that moisture regain values decreased slowly but progressively with strength of the dose, but at 100 Mrad a small but definite rise in moisture regain was observed. The rise in regain value in the case of purified cotton at the same dose has also been reported by Blouin and Arthur [25].

5.1 Control of Moisture during Processing of Jute Fiber

Jute, being a very coarse fiber, cannot be economically processed in a dry stage, otherwise there will be tremendous loss due

5 MOISTURE IN JUTE AND SIMILAR FIBERS

to wastage, deterioration of quality resulting in the production of bad, irregular, nonuniform, and hairy yarns. The moisture regain in exloom cloth has to be maintained at 13 to 15%. For the judicious control of the moisture and better spinnability and consequent fabrication of the jute cloth, the control over atmospheric conditions is most desirable. The optimum limits of moisture in the atmosphere at different stages of jute processing range from 60 to 85% [145]. The fibers at different processing stages contain different moisture regain values. Approximate average moisture regain percentage values at different processing stages are shown in Table 9.8. There is thus absorption or desorption of the moisture by jute fibers [146], depending on the variation in relative humidity as is the case with cotton and wool fibers.

Jute is very coarse; therefore, to render it pliable and economically processable, water is added along with oil in the form of an emulsion on the softening machines. Though water and oil are immersible, they are formed into an emulsion by the addition of a small quantity of emulsifying agent. The emulsion is formed in oil

TABLE 9.8

The Regain Values in Different Stages of Jute Production

Sample no.	Stage of production	Regain (%)
1	Breaker card	34 to 36
2	Finisher card	30 to 32
3	Drawing frames	26 to 29
4	Spinning	17 to 18
5	Weaving	20 to 25
6	Finishing	
	Hessian	16 to 17
	Sacking	17 to 19

in the water phase by dissolving the emulsifying agent in water. The real softening agent of the fiber is water, whereas the function of the oil is to check the evaporation of water from the fiber and to reduce friction between machine and the fiber. What should be the standard percentage of the mixing of oil and water is now known, but generally, the emulsion applied on the jute fiber is 25%. The application of oil for better spinning performance, as recommended by IJMARI (Indian Jute Mills Association Research Institute), is 5 to 6% calculated on the dry weight basis of the fiber. The amount of water that should be added along with oil cannot be rigidly maintained as it depends on atmospheric conditions which control the rate of evaporation, which may well be the deciding factor.

It is desirable that the jute contain a suitable percentage of moisture, for if the fibers are too wet, they are easily pulled apart, and easily stretched beyond their elastic limit and hence weakened. Generally 16 to 20% of water is added in the form of an emulsion on the jute in a softening machine. Required adjustment in decrease or increase of the amount of water may be made, depending on the variation of relative humidity with changing season.

Size is applied to jute warp yarns with the main object of laying the surface hairs by means of an adhesive coating and thus enabling the yarn to better withstand the abrasion effects of the weaving operation. Since the size solutions are applied from liquid medium, the sized yarn contains an excess of water which has to be removed in the subsequent drying operation, and the moisture content of the sized yarn after drying is important. Excessive drying will give brittle yarns, and also reduces weight because of more moisture evaporation, while on the other hand, a damp yarn is liable to develop mildew strains which can eventually deteriorate the yarn. So it is most important to control the moisture content of the sized yarn so that the over- or underdrying of the warp is avoided and the correct moisture content is maintained resulting in a good weaving efficiency and at the same time checking the mildew formation.

5 MOISTURE IN JUTE AND SIMILAR FIBERS 647

The percentage of the moisture content to be maintained on the weaver's beam after sizing depends on moisture evaporation due to changing relative humidity. Usually it is the practice of every mill to apply more moisture on sacking beams than on hessian. Approximately 20 to 25% moisture regain is expected to give good results in moderate seasons. A slight variation may be made with fluctuating relative humidity. A little addition of oil to the sizing mixture may prove to be beneficial so as to check moisture evaporation to some extent and to impart a little pliability to the yarn for better weaving operation.

The standard regain and moisture content of jute at 65% relative humidity and 70°F are 13.75 and 12.09% respectively. Saturation limit or the saturation regain is the maximum amount of moisture fiber can hold at 100% relative humidity, for jute it is approximately 35%. The final stage of the application of the water on the jute processing is the addition of water on the damping machine. The process of damping is an intermittent stage between the weaving shed and calender machines. This addition of water on the damping machine has a specific importance. When the cloth is woven on the looms, because of its mechanical processing and the cloth being held under tension till the completion of the weaving of the full cut, stress is developed in the fibers. Calendering of the cloth is the essential process to render the fabric a good cover by flattening the yarn and imparting glaze. For glaze, cloth passes round the steam roller where the excessive moisture is evaporated off and the regain brought down to safe limits.

To release the stress developed during weaving and to compensate for the moisture evaporated during calendering, the cloth is passed over the damping machine where a fine spray of water is deposited on the cloth surface. The control is exercised to put in the correct amount of moisture. The spraying nozzles should be kept clear of clogging dirt and dust, otherwise irregular damping will result and this causes bias on cloth due to dry-part slipping and the damp parts gripping the tension rails. Generally water applied in damping is 3 to 5%; application of a greater or lesser quantity of water than the actual requirement affects the cloth as on this application depends the moisture regain in the finished cloth. Variation of moisture in damping also affects to some extent the increase in the length of finished cloth and hence its width. In the finished cloth, 16 to 18% moisture regain in hessian and 17 to 19% in sacking are usually maintained. Higher regains would produce mildew in the bales and reduction of cloth strength. Though the correct moisture regain of the bales depends on evaporation, it should be maintained slightly above the safe limit so that the natural evaporation would bring the regain within safe limits.

5.2 Methods of Moisture Measurement

The usual methods and techniques for the measurement of moisture content of textile materials described earlier are equally applicable for the measurement of the noisture content of loose jute as well. Banerjee and Sen [147] have developed a special resistance or conductivity-type moisture meter for the measurement of moisture content of loose jute fibers and jute bales. This electronic moisture meter has been described in their publication; however, a short description of this instrument as applied to jute bales will be given here.

In the case of jute bales, the electrodes are a pair of stout steel rods each circular in cross section and gradually tapering to a pointed end. The total length of the electrode is 18.5 in. There is a binding screw in each for making electrical connections to the meter. The positions of the electrodes marked on a jute bale are shown in Fig. 9.18.

Three readings for each bale are taken by introducing the electrode successively into the positions marked A_1B_1, A_2B_2, and A_3B_3 (Fig. 9.18). Gentle hammering on the electrodes is necessary to ensure tight contact with fibers. The average of the three readings with the electrodes in the positions marked as in Fig. 9.18 has been found from many preliminary trials to give the representative value

5 MOISTURE IN JUTE AND SIMILAR FIBERS   649

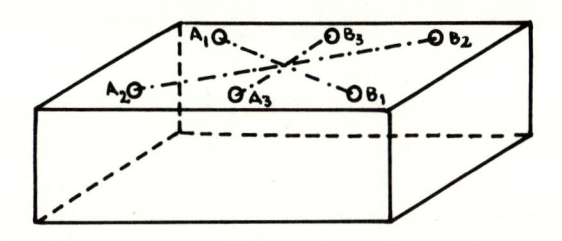

FIG. 9.18. Electrode system of IJMARI moisture meter for jute bales.

of the percent moisture regain of the whole bale. The bale is unpacked and samples from its different parts are tested with the same instrument using the electrode for loose jute in the manner described earlier. In all, 25 readings of the meter are taken for each bale and the average moisture content is obtained. Nearly 100 bales were tested in this way and the resulting calibration curves are shown by the full lines in Fig. 9.19. It will be observed that the broken and the full-line curves almost coincide with each other, indicating that the same calibration curves may serve the purpose of determining the moisture contents of jute bales as well as of loose jute.

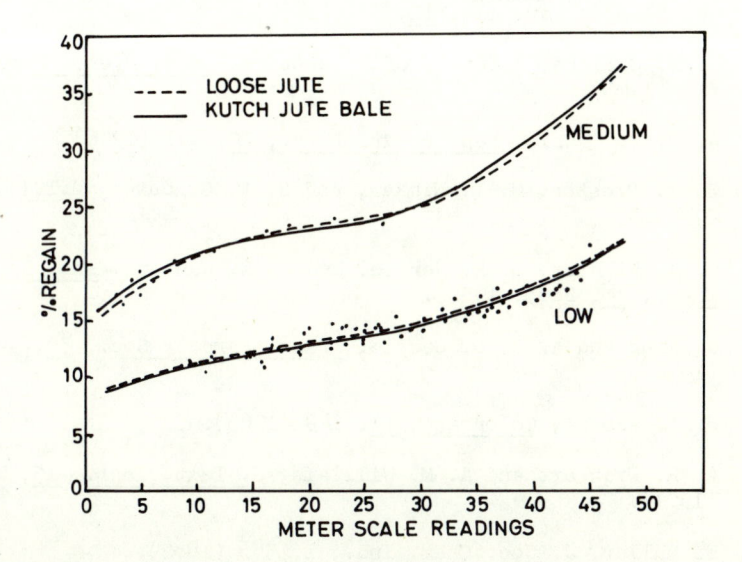

FIG. 9.19. Calibration curve for jute.

REFERENCES

1. Anonymous, Determination of Regain in Wool, *Wool Sci. Rev.*, 19, 21 (1961).

2. R. K. Toner, C. F. Bowen, and J. C. Whitewell, *Text. Res. J.*, 18, 526 (1948).

3. R. K. Toner, C. F. Bowen, and J. C. Whitewell, *Ibid.*, Part II, 19, 1 (1949).

4. J. C. Whitewell and R. K. Toner, *Ibid.*, Part III, 19, 755 (1949).

5. J. C. Whitewell, C. F. Bowen, and R. K. Toner, *Ibid.*, Part IV, 20, 400 (1950).

6. A. Pande, Moisture in Textiles, *Text. Rec.*, Part I, 963, 70 (1963); Part II, 964, 54 (1963); Part III, 965, 55 (1963); Part IV, 966, 103 (1963).

7. A. Pande, Moisture in Textiles, *Ibid.*, Part V, 967, 105 (1963); Part VI, 968, 97 (1963); Part VII, 969, 62 (1963).

8. A. Pande, Moisture in Textiles, *Ibid.*, Part VIII, 970, 46 (1964); Part IX, 971, 67 (1964); Part X, 972, 55 (1964).

9. A. Pande, *Lab. Pract*, Part I, 12, 432 (1963); Part II, 12, 661 (1963); Part III, 12, 741 (1963).

10. P. W. Carlene, A Survey of Literature, *Amer. Dyest. Rep.*, 34, 322 (1945).

11. A. B. D. Cassie, *Rep. Progr. Phys.*, 10, 141 (1946).

12. J. M. Preston, M. V. Nimkar, and S. P. Gundavda, *J.T.I.*, 42, T79 (1951).

13. G. A. Richter, L. E. Herdle, and W. E. Wahtera, *Ind. Eng. Chem.*, 44, 2883 (1952).

14. G. King and A. B. D. Cassie, *Trans. Faraday Soc.*, 36, 445 (1940).

15. W. B. Achwal, *Colourage*, 11, 149 (1965).

16. A. R. Urquhart and A. M. Williams, *J. Text. Inst.*, 15, T138 (1924).

17. E. Müller, *J. Soc. Chem. Ind.*, 1, 356 (1882).

REFERENCES

18. R. L. N. Iyengar and J. Prakash, ISI Bull., 12, 227 (1960).

19. Textile World Yearbook, McGraw-Hill, New York, 1938, p. 38.

20. A. R. Urquhart and A. M. Williams, J. Text. Inst., 15, T559 (1924).

21. J. G. Wiegerink, J. Res. Nat. Bur. Std., 24, 645 (1940).

22. F. S. Perkerson, W. A. Reeves, and V. W. Tripp, Text. Res. J., 30, 179 (1960).

23. W. A. Reeves, R. M. Perkins, and L. H. Chance, Ibid., 30, 179 (1960).

24. J. A. Howsman, Ibid., 19, 152 (1949).

25. F. A. Blouin and J. C. Arthur, Jr., Ibid., 28, 198 (1958).

26. R. A. Gill and R. Steele, Ibid., 32, 338 (1962).

27. A. Pande, B. Singh, C. L. Jain, and V. B. Chipalkatti, Ibid., 39, 1117 (1969).

28. TAPPI Standard T 412-sm-1969.

29. ASTM Designation D644T and ASTM Standard 1942, Part III, pp. 10004-10006, and 1944, Part II, pp. 341-343.

30. T. J. Mitchell, Chem. Ind., 751 (1950).

31. K. Göhde, Papier Fabr., 37, 320 (1939).

32. C. Duval, Inorganic Thermogravimetric Analysis, Elsevier, Amsterdam, 1953.

33. S. G. Barker and J. J. Hedges, Wool Ind. Res. Assoc. Publ., 71, 13 (1941).

34. J. G. Downes, J. Sci. Instrum., 38, 484 (1961).

35. R. B. Brock, Chem. Ind., 299 (1947).

36. E. Hummel, Text. Pra. (Int. Engl. Ed.), 38 (1958).

37. W. E. Kuntzal, J. Amer. Chem. Soc., 61, 514 (1939).

38. A. Zimmerman, Fette, Seife., 46, 446 (1939).

39. J. Mitchell, Jr., J. Ind. Eng. Chem., 12, 390 (1940).

40. J. F. Keating and W. M. Scott, Amer. Dyest. Rep., 31, 308 (1942).

41. A. Pande and C. S. Pande, Instrum. Pract., Part I, 16, 896 (1962); Part II, 16, 988 (1962); Part III, 16, 1104 (1962).

42. E. H. Jones, J. Sci. Instrum., 17, 55 (1940); British Pat. 516,379 (1940).

43. J. A. Reddick, S. C. Myne, Jr., and E. E. Berkley, Text. Res. J., 29, 220 (1959).

44. J. E. Fielden, British Pat. 619,534 (1954); Electron Eng., 21, 10 (1949).

45. A. Pande, Instrum. Pract., 19, 650 (1965); Indian Pat. 97,107.

46. Anonymous, Amer. Dyest. Rep., 45, 935 (1956).

47. I. C. Watt and J. D. Leeder, Trans. Faraday Soc., 60, 1335 (1964); J. Soc. Dyers Color., 83, 226 (1967).

48. K. Baird and P. L. Fitz, Text. J. Aust., 44, 16 (1969).

49. A. R. Urquhart and N. Eckersall, J. Text. Inst., 21, T510 (1930).

50. K. Biltz, Text. Forschungber., 3, 89 (1921).

51. E. A. Hutton and J. Gartside, J. Text. Inst., 40, T161 (1949).

52. E. A. Hutton and J. Gartside, Ibid., 40, T170 (1949).

53. M. V. Forward and S. Smith, J. Text. Inst., 46, T158 (1955).

54. A. A. Armstrong, Jr. and W. K. Walsh, Report prepared for the Division of Isotopes Development, U.S. Atomic Energy Commission, October 31, 1962.

55. A. Angster and M. Wandel, Milliand Textilber, 38, 1411 (1957).

56. A. Elliot, W. E. Hanley, and B. R. Malcon, Brit. J. Appl. Phys., 5, 377 (1954).

57. E. R. S. Jones, J. Sci. Instrum., 30, 132 (1953).

58. H. H. Schenker, Anal. Chem., 29, 825 (1957).

59. J. Haslam and M. Claser, Analyst, 413 (1950).

60. Lyons, Proc. Amer. Phys. Soc., 80, 802 (1946).

REFERENCES

61. H. H. Walker, Text. Res. J., 28, 1047 (1958).

62. S. A. Shorter and W. J. Hall, J. Text. Inst., 15, T305 (1924).

63. J. B. Speakman, J. Soc. Chem. Ind., 49, T209 (1930).

64. A. C. Goodings, Amer. Dyest. Rep., 24, 109 (1935).

65. D. K. Ashpole, Proc. Roy. Soc. (London), 212A, 112 (1952).

66. I. C. Watt and R. H. Kennet, Text. Res. J., 30, 489 (1960).

67. M. Feughelman and I. C. Watt, Ibid., 31, 960 (1961).

68. J. B. Speakman and C. A. Cooper, J. Text. Inst., Part I, 27, T183 (1936); Part III, 27, T191 (1936).

69. M. Harris, J. Res. Nat. Bur. Std., 29, 73 (1942).

70. E. F. Mellon, A. H. Korn, and S. R. Hoover, J. Amer. Chem. Soc., 71, 2761 (1949).

71. C. H. Nicholls and J. B. Speakman, J. Text. Inst., 45, T267 (1954).

72. E. G. H. Carter, W. R. Middlebrook, and H. Philips, J. Soc. Dyers Color., 62, 203 (1946).

73. I. C. Watt, Text. Res. J., 33, 631 (1963).

74. I. C. Watt, Ibid., 30, 443 (1960).

75. A. B. D. Cassie, Trans. Faraday Soc., 36, 453 (1940).

76. G. B. MacMohan and J. G. Downes, Intern. J. Heat and Mass Transfer, 5, 689 (1962).

77. P. Nordon, Nature, 200, 1065 (1963); Intern. J. Heat and Mass Transfer, 7, 639 (1964).

78. P. S. H. Henry, Proc. Roy. Soc., A171, 215 (1939); Discuss. Faraday Soc., 3, 243 (1948).

79. A. Edmunds, N.Z. J. Sci., 4, 775 (1961).

80. I. K. Walker, Ibid., 6, 127 (1963).

81. P. Nordon, N. W. Bainbridge, and A. Edmunds, Wool Technol. Sheepbreeding, 11, 51 (1964).

82. W. Von Bergen and C. S. Clutz, Text. Res. J., 20, 580 (1950).

83. M. Cednas, J. Text. Inst., 52, T251 (1961).

84. K. Baird, Text. Res. J., 33, 973 (1963).

85. K. Baird, Ibid., 31, 1038 (1961).

86. F. Bekku, J. Soc. Text. Cellul. Ind. (Japan), 14, 631 (1958).

87. T. S. Nutting, J. Text. Inst., 52, T407 (1961).

88. B. Olofsson, Ibid., 52, T272 (1961).

89. B. Olofsson, Text. Res. J., 32, 1037 (1962).

90. A. R. Haly and J. W. Snaith, Ibid., 33, 872 (1963); 34, 1 (1964); 35, 147 (1965).

91. M. Feughelman, Ibid., 29, 223 (1959).

92. M. Feughelman, A. R. Haly, and J. W. Snaith, Ibid., 32, 913 (1962).

93. W. Von Bergen, Ibid., 29, 586 (1959); Ind. Eng. Chem., 44, 2157 (1952).

94. F. T. Pierce, J. Text. Inst., 20, T133 (1929).

95. A. B. D. Cassie, Fibrous Proteins Proc. Symposium Soc. Dyers Color., 86 (1946).

96. G. A. Gilbert, Ibid., 96 (1946).

97. J. R. Katz, Trans. Faraday Soc., 29, 279 (1933).

98. S. Brunauer, Physical Adsorption, Princeton Univ. Press, Princeton, N. J., 1943; S. Brunauer, P. H. Emmett, and E. Teller, J. Amer. Chem. Soc., 60, 309 (1938).

99. A. B. D. Cassie, Trans. Faraday Soc., 41, 450 (1945).

100. T. L. Hill, J. Chem. Phys., 14, 263 (1946).

101. A. D. McLaren and J. W. Rowen, J. Polym. Sci., 7, 289 (1951).

102. A. J. Hailwood and S. Horrobin, Trans. Faraday Soc., 42B, 84 (1946).

103. J. W. Rowen and R. Simha, J. Phys. Colloid Chem., 53, 921 (1949).

104. P. J. Flory, J. Chem. Phys., 10, 51 (1942).

REFERENCES

105. M. L. Huggins, Ann. N. Y. Acad. Sci., 32, 1 (1942).

106. A. D. Sule, Wool Woollens India, 3, 23 (1967).

107. J. J. Windle and T. M. Shaw, J. Chem. Phys., 25, 435 (1958).

108. M. Feughelman and A. R. Haly, Text. Res. J., 32, 227 (1962).

109. E. F. Mellon, A. H. Korn, and S. R. Hoover, J. Amer. Chem. Soc., 69, 827 (1947).

110. J. E. Moore, Text. Res. J., 26, 936 (1956).

111. A. R. Haly and J. W. Snaith, Ibid., 31, 983 (1961).

112. P. H. Hermans, Physics and Chemistry of Cellulose, Elsevier, Amsterdam, 1946, 1949.

113. Jordan, Lloyd, and Phillips, Trans. Faraday Soc., 30, 382 (1934).

114. Jordan, Lloyd, and Moran, Proc. Roy. Soc., A147, 342 (1934).

115. A. R. Haly, Sonderdruck Kolloid Z. Z. Polymer, 189, 43 (1963).

116. J. L. Morrison, Nature, 198, 84 (1963).

117. J. G. Downes and P. Nordon, Ibid., 203, 719 (1964).

118. I. C. Watt, Text. Res. J., Part II, 30, 644 (1960).

119. P. Mason, Ibid., 34, 747 (1964).

120. M. Feughelman, Ibid., 29, 967 (1959).

121. I. C. Watt, R. H. Kennet, and J. F. P. James, Ibid., 29, 975 (1959).

122. M. Feughelman and I. C. Watt, Ibid., Part IV, 31, 962 (1961).

123. H. P. Rothbaum and V. N. M. Vera-Jones, Text. J. Aus., 35, 932 (1960).

124. B. H. Mackay, Text. Res. J., 32, 81 (1962).

125. International Wool Textile Organization, International Regulations for the Conditioning of Wool (1954); International Standards Organization Recommendation R. 139.

126. British Standards Institute, Brit. Stand. 1051 (1953).

127. B. H. Mackay, *J. Text. Inst.*, 54, T376 (1963).

128. B. H. Mackay, J. G. Downes, and I. C. Watt, *Ibid*, 54, T350 (1963).

129. I. C. Watt, R. H. Kennet, and J. G. Downes, *Ibid*, 52, T46 (1961).

130. G. C. LeCompte and H. H. Lipp, *Amer. Dyest. Rep.*, 38, 484 (1949).

131. E. W. Dean and D. D. Stark, *Ind. Eng. Chem.*, 12, 486 (1920).

132. K. Fischer, *Angew. Chem.*, 48, 394 (1935).

133. J. R. Wilson and M. M. Sandmoire, *Text. Res. J.*, 30, 587 (1960).

134. ASTM Method of Test for Moisture in Wool by Oven Drying Designation D 1576-58T, Amer. Soc. Test. Mater., Part III 4(a), 1944 (1952).

135. N. C. Mitra and J. Venkataraman, *J. Soc. Chem. Ind.*, 57, 306 (1938).

136. A. Pande, *Modern Hygrometry*, Somaiya Publ. Bombay, India, 1970.

137. J. Lemann, *Naturewissenschaften*, 45, 35 (1958).

138. N. F. Roberts and J. E. Algie, *J. Text. Inst.*, 54, T464 (1963).

139. J. E. Algie, *Text. Res. J.*, Parts I and II, 36, 1026 (1964); 317 (1964); *Kolloid Z.*, 234, 1069-1078 (1969).

140. A. Pande, Unpublished results (1964).

141. R. M. Shaw and R. H. Elsken, *J. Agr. Food Chem.*, 27, 1983 (1955).

142. G. W. West, A. R. Haly, and M. Feughelman, *Text. Res. J.*, 31, 899 (1961).

143. R. V. Pound and W. D. Knight, *Rev. Sci. Instrum.*, 21, 219 (1950).

144. S. K. Majumdar and W. H. Rapson, *Text. Res. J.*, 33, 1 (1963).

145. G. Rukhana, *Text. Trends (India)*, 65 (1964).

146. H. Chatterjee, *J.S.I.R. (India)*, 20D, 286 (1961).

147. B. L. Banerjee and M. K. Sen, *J.S.I.R. (India)*, 15A, 24-28 (1956).

Chapter X

MOISTURE IN BAGASSE, WOOD, AND PAPER

1 MOISTURE IN BAGASSE

1.1 Introduction

Bagasse consists of spongy fibers, in which is absorbed extremely diluted cane juice, the two being present in about equal quantities by weight. In modern factories the following may be taken as the average composition of bagasse by weight:

Fiber	46 to 49%
Water	45 to 48%
Sugar	2.5%
Nonsugars	0.5%

The presence of water in such a large proportion naturally makes bagasse an unusually low-grade fuel, with characteristics so different from other industrial fuels that a special type of furnace is necessary for satisfactory combustion.

The first large-scale industrial application of bagasse was in the manufacture of insulating board, the development of Calotex has been reviewed by Lathrop [1]. Canex is another example of its application in industry for which the cellulose and hemicellulosic material of the whole fiber is employed. Lynch and Goss [2] have conducted numerous investigations of the possibilities of producing high-grade α-cellulose chemical pulp economically from bagasse. The nitric acid digestion method which they developed is in industrial use in some countries. A process was later evolved from their studies which utilizes principally the hemicellulose and lignin of bagasse for the manufacture of molding plastics. The development of this product, and its application in the making of phonograph records and other molded products has been described by Molitor [3]. Bagasse is used to some extent in the manufacture of corrugating board, and the possibility of using it wholly or partially for making newsprint paper has received attention from time to time.

Moisture plays a very important role in the preservation and utilization of bagasse. It is, therefore, necessary to have two types of measuring devices for the measurement of moisture content of bagasse in bulk as well as in continuous processes. A few methods developed for the measurement of moisture content of bagasse are described. It is, however, assumed that the other methods and techniques which have been described earlier for the measurement of moisture content of fibrous materials such as cellulose, jute, wool, etc., are equally applicable for the measurement of the moisture content of bagasse.

1.2 Methods of Moisture Measurement

There exists a large amount of unpublished analytical data on bagasse. Considering the large number of varieties of cane, as well as the many different localities from which it is harvested, and the nature of the material, there is exhibited good agreement in the published analytical results. From the analytical results available in literature, it is safe to assume that the cellulose content of moisture-free bagasse averages one half its weight. Bagasse in equilibrium with ambient air has been reported to contain 8.3% of

1 MOISTURE IN BAGASSE

moisture by Kumagawa and Shimomura [4], 10.8% moisture by the Imperial Institute Laboratory [5], 10.3% water by Valenzuela and West [6]. Lynch and Goss [2] have reported that random samples of bagasse contain from 9.95 to 12.58% moisture.

Besides the standard oven-drying method, an electrical method based on the dielectric properties of bagasse-water mixtures has been used by Munro and Wise [7] for the measurement of moisture content of bagasse having different degrees of moisture contents. They used for their experiments a commercial electrical moisture meter known as the Kappa Moisture Meter (manufactured by Kappa Moisture Meters Ltd., England). This meter measures the dielectric constant of a sample containing moisture by the frequency change which takes place when the latter is introduced into a radiofrequency circuit.

For a detailed description of dielectric as well as conductivity types of moisture meters, the reader is referred to a review publication [8] and Chap. V, where detailed design features and constructional details of different types of electrical and electronic moisture meters are described and their merits and demerits have been discussed.

One of the chief difficulties in the determination of moisture in bagasse is the difficulty of obtaining a representative sample which is small enough to be conveniently handled. The sample cell of the Kappa moisture meter used by Munro and Wise [7] had a capacity of about 400 g of wet bagasse. With such large samples of bagasse, however, the adjustment of the meter to the balance point becomes very difficult owing to a great reduction in the intensity of the radiofrequency used. It was, therefore, found necessary to use samples weighing 100 g or so. In the original experiments of Munro and Wise [7] a standard weight of 100 g was chosen so as to be able to obtain a readable balance point on the electrical meter. In order to study the effect of sampling, a few experiments were made in which a larger sample of 300 g of partly dried bagasse was successfully used. In these experiments a constant drying time of 20 min was adopted. It was found that with 300-g samples the standard error was 1.2% whereas the standard error in 100-g samples was 2.5%. There is, therefore, a considerable advantage in using a larger sample than 100 g in the instrument. By extending the drying time slightly, even better accuracy could probably be obtained.

Another problem connected with moisture measurement of bagasse is that samples tend to dry out on being handled for even short periods of time in the atmosphere. This effect is considered likely to be of special importance when using the dielectric moisture meter since it might well be desirable to obtain replicate meter readings on the same sample. In order to be able to estimate the standard error of the meter readings on each sample, 10 readings were made on each fraction. The variation of standard error with number of observations is shown in Fig. 10.1. This standard error includes a contribution due to evaporative losses as well.

A combination of oven drying and dielectric method was used by Munro and Wise [7] for determination of high moisture content values of bagasse. The bagasse, after being dried in the vacuum oven, was also tested in the moisture meter. The average meter reading

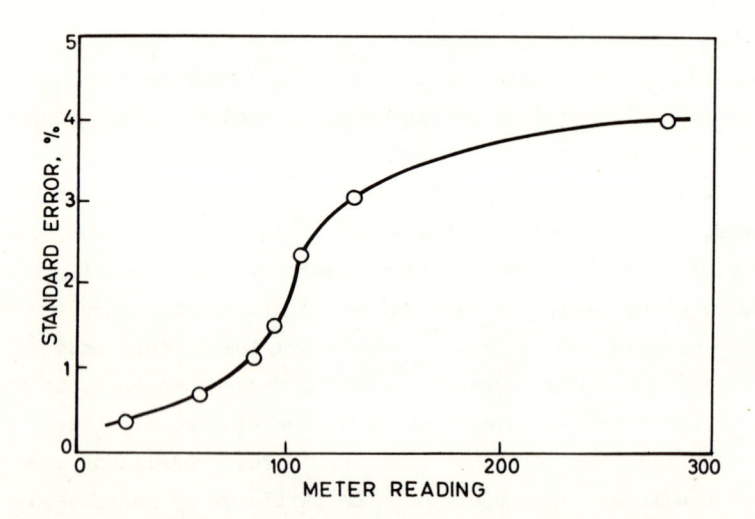

FIG. 10.1. Graph showing the variation of the standard error of measurement with meter readings of Kappa moisture meter.

obtained was 23.9 with a standard error of 0.6; this suggests that the meter becomes more accurate with the dried bagasse and leads to a possible procedure whereby the bagasse sample is partly dried in an oven and then the residual moisture is determined with the electrical moisture meter.

This combined method of partly oven drying and partly using the electronic moisture meter is a compromise between the two. However, it takes much less time than oven drying alone and gives more accurate results as compared with the electrical method. Also samples larger than 100 g can be used with the meter if the bagasse is partly dried. This procedure solves one of the chief difficulties in the method of determining the moisture in bagasse, i.e., having a larger representative sample which can be conveniently handled. A calibration curve (Fig. 10.2) was obtained between the Kappa moisture meter readings and moisture content values of bagasse by this procedure.

Munro and Wise [7] have compared the results obtained by the electrical method to those by Dean and Stark's [9] distillation method. The standard error of the Dean and Stark method was found to be 1.5%. This includes a sampling error. As in the Dean-Stark apparatus a smaller sample of about 50 g only can be employed, the error is approximately 1%. A sampling error of 1% would actually give a standard error for a single Dean and Stark determination also of 1%.

As to the rapidity, the Dean and Stark [9] and the combined oven-drying and electrical methods are roughly equivalent, both requiring just less than 1 hr. However, by careful examination of the technique used, both methods can be made slightly faster. According to Munro and Wise [7] the electrical method is the superior one, requiring no fragile glassware or flammable solvents. The former is applicable to other sugar products which the sugar factory laboratories may well have to investigate from time to time. It therefore appears reasonable to conclude that dielectric methods of

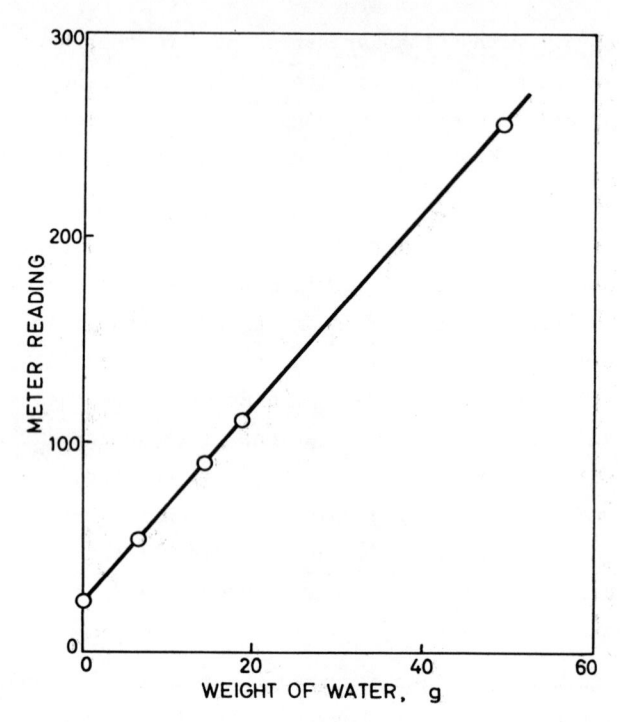

FIG. 10.2. Calibration curve between the readings of Kappa moisture meter and moisture contents of bagasse samples.

moisture determinations are likely to be of considerable value to the sugar industry.

The chief difficulty of measuring (by electrical methods) higher moisture contents which are found in the case of bagasse (as coming out of the last crushing stages of the sugar mill) is the increased conductance of the sample due to the presence of an excessive amount of free water which cannot be absorbed by the hygroscopic cellulosic material of bagasse during the process of secondary extraction of juice from the sugar cane. Munro and Wise [7] have observed that dielectric loss values of bagasse depend largely on conductance of the sample and suggested that this effect can be eliminated by using high-frequency fields. However, Pande and Aggarwal [10] have recently found that the effect of high-frequency

1 MOISTURE IN BAGASSE

fields is not appreciable. A number of other factors such as temperature, bulk or packing density, mechanical distribution of water within the sample, and particle size and shape of the bagasse samples complicate the accurate moisture measurement by the dielectric method.

These difficulties have been overcome by designing a special electrode system in which the metal plates of the electrode of the dielectric moisture meter have been coated with a high polymeric material such as methacrylate, polystyrene, or polyester. It was found that coating of the electrode plates with these materials considerably reduces the conductance and leakage current, which enables the electrode to behave as a capacitor. A mechanism has also been provided in the electrode system due to which the sample is kept at a predetermined constant pressure so that the same packing density is maintained at which the moisture meter was calibrated. A schematic diagram of this electrode system is shown in Fig. 10.3. A series of experiments have been conducted on bagasse samples having moisture contents as high as 50% on wet weight basis and a linear relationship between moisture values and electrical currents has been obtained. The results obtained by the new electrode system in combination with the SRI electronic moisture meter are in good agreement with those obtained with ordinary oven drying as well as vacuum oven-drying methods [10].

The operational studies of this moisture meter were conducted [10] in an experimental sugar mill where the following experiments were performed:

    1. The moisture content of composite bagasse was determined at the same time by oven drying and electronic methods using the same sample.

    2. The particle size of the sample of composite bagasse was varied by sieving through 1, 4, 10, and 12 mesh and then the moisture was determined by vacuum oven-drying method and electronic method at the same time.

    3. The moisture of the composite bagasse was artificially varied and the moisture was determined by both the methods again.

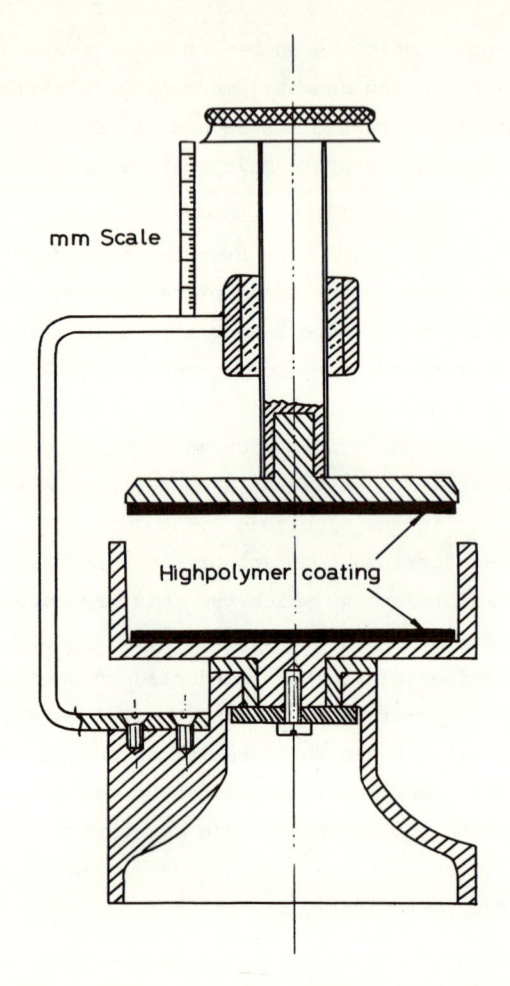

FIG. 10.3. Schematic diagram of the electrode system for bagasse.

4. The composite bagasse was sieved as in case 2 and the moisture was varied artificially in each case. The moisture was determined by both the methods as before.

A number of experiments were performed in each case and the data obtained are given in Tables 10.1 and 10.2. It is found that the results of moisture meter against oven-drying methods in cases 1 and 3 are satisfactory within experimental limits. In experiments

1 MOISTURE IN BAGASSE

TABLE 10.1

Results of Moisture Measurement of Composite Bagasse
(Comparison of SRI Electronic Moisture Meter with
Ordinary Oven and Vacuum Drying Methods)

Sample no.	Moisture in bagasse by oven-drying method (%) Under vacuum	Ordinary	Moisture by electronic moisture meter (%)
1	42.78	42.80	44.50
2	44.20	45.24	45.30
3	43.18	33.58	44.50
4	42.79	45.04	45.50
5	39.20	41.06	42.00
6	49.21	48.70	48.00
7	42.98	43.10	42.30
8	48.32	47.97	46.60
9	45.37	44.98	45.50
10	47.54	46.48	46.85

TABLE 10.2

Variation of Moisture Content with Variation of Particle Size of
Bagasse (Composite Bagasse Sieved) as Determined by SRI Moisture
Meter and Ordinary Oven and Vacuum Drying Methods

Size of sample particles (mesh)	Moisture in bagasse by oven-drying method (%) Under vacuum	Ordinary	Moisture by electronic moisture meter (%)
1	44.86	45.59	43.0
4	44.64	46.81	44.5
10	45.93	46.55	49.0
12	44.24	43.31	51.0

of case 1, the standard deviation calculated from the data is 0.206. In case 3, the results of the moisture meter are quite satisfactory up to about 55% moisture on the wet weight basis. In cases 2 and 4, considerable discrepancy in the moisture meter and oven-drying results was observed as was expected. The results of moisture meter are satisfactory for composite bagasse, for which it was calibrated. However, it is observed that greater sensitivity of the electronic circuit of the moisture meter is required in the moisture range 45 to 55%. This is the range which is of interest to the sugar industry for the measurement of moisture content of bagasse. It may, however, be pointed out that the requirement of the sugar industry for the measurement of moisture content is that of composite bagasse only. The readings of sieved bagasse having different particle sizes were taken for academic interest only, i.e., to find out the correlation between particle size and moisture content and to evaluate the validity of calibration on composite samples of bagasse to fractionated constituents. The effect of packing density of bagasse on the variation of the dielectric constant of the bagasse-water mixture has also been studied [10] and it has been found that size and shape factors have appreciable effect on the dielectric constant values of the bagasse-water mixture.

2 MOISTURE IN BEET PULP

2.1 Introduction

Beet pulp, the byproduct from the diffusion of sugar from beet cossettes during sugar manufacture, comes from the diffusion battery with a moisture content of about 90%. Usual practice involves dewatering over a screen and, where dehydration of the pulp is practiced, final pressing is done to reduce the moisture content to 80 to 86%. Old presses, or presses improperly adjusted, deliver pulp having moisture contents as high as 90%. Where several presses operate to remove water from beet pulp prior to dehydration, a rapid

2 MOISTURE IN BEET PULP

means of evaluating individual press performance is needed. A press which one day produces pulp having 85% moisture may give pulp containing 88% the next day. This corresponds to a change of from 5.66 to 7.32 in the ratio of water to dry matter. In other words, the amount of water to be evaporated per pound of dry matter increases 29% when pressed pulp moisture goes from 85 to 88%.

2.2 Methods of Measurement

Cotton et al. [11] have developed a simple control method enabling operators to determine moisture in approximately 6 min. They have also given data on the correlation between this rapid method of moisture measurement and the oven-dry method. They have also compared the reproducibilities of the two methods. The principle of this analytical method is the release of moisture from pressed pulp by contact with molasses. This is followed by a final pressing in a simple potato ricer. The volume of liquid released by the ricer is directly related to the moisture content of the pulp.

The basic unit of the apparatus is a household potato ricer (Fig. 10.4) consisting of a perforated basket, a plunger with attached lever, and fulcrum. Attached to the end of the lever is a 20-lb weight so that constant pressure can be applied during each moisture determination. The ricer is mounted on a stand. A funnel below the screen leads the press liquor into a graduated cylinder. The procedure to be followed is: Weigh 200 g of pressed beet pulp into a beaker and weigh into the beaker 200 g of molasses at room temperature. Mix manually with a V-shaped wire for 1 min. Place contents of the beaker in the potato ricer (Fig. 10.4). Press for exactly 2 min, using the weight as a source of pressure. Catch the press liquor in a 250-ml graduated cylinder. Record volume of press liquor. Read moisture content of pulp from a plot of press volume versus moisture content (Fig. 10.5).

Figure 10.5 represents data obtained from 95 determinations of press-liquor volume vs moisture determined by vacuum-oven drying at

668   X.   MOISTURE IN BAGASSE, WOOD, AND PAPER

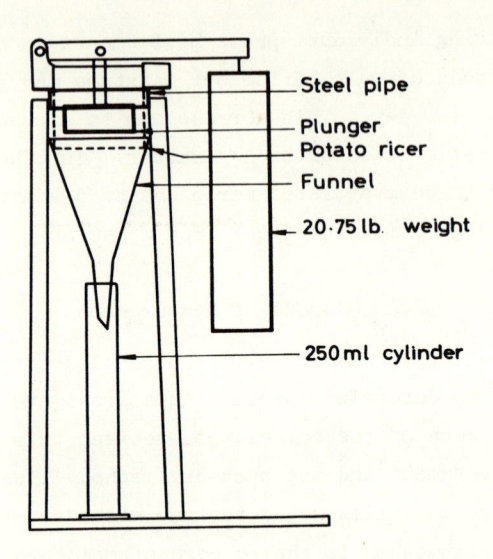

FIG. 10.4. Apparatus for rapid determination of moisture of beet pulp.

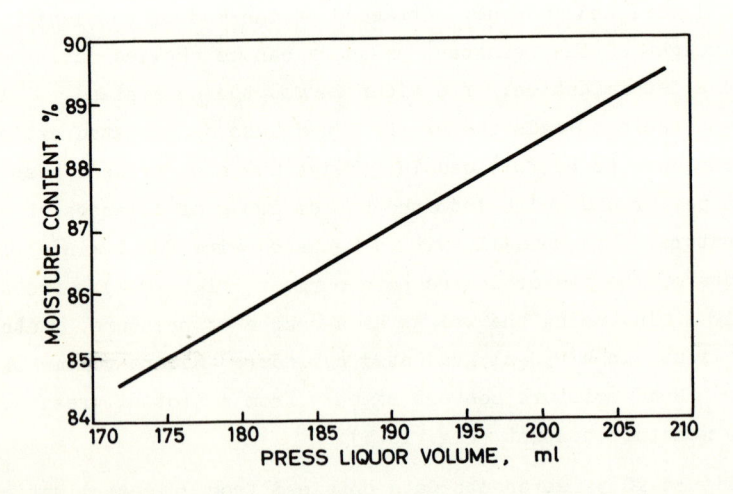

FIG. 10.5. Graph showing the moisture content of beet pulp versus press volume.

100°C to a constant weight. The coefficient of correlation between press liquor volume and percent moisture obtained from the 95 comparisons was calculated to be 0.9664, which is a very high degree of correlation. The curve in Fig. 10.5 is based on determinations made during a week in a factory. Once a correlation is established the data can be analyzed in order to compare the precision of the new method with the oven-dry procedure. Table 10.3 gives standard deviations from 10 experiments, comparing press volume with oven-dry moisture determinations; each experiment consisted of from 4 to 12 replicate determinations on a given sample of pulp. The average of the standard deviations for moisture content by the two methods was approximately the same, that is, 0.1594 and 0.1489. The rapid method is therefore as precise as the oven-dry procedure. It has the advantage of speed. Furthermore, a foreman can use it right in the pulp dryer building and immediately spot a press that is improperly adjusted. Only occasional checks by a chemist are needed to be sure the method is behaving satisfactorily. It was observed by Cotton et al. [11] that temperature variations between 23 and 40°C did not appreciably affect the precision of the method.

3 MOISTURE IN WOOD AND TIMBER

3.1 Introduction

The absorption of water by wood belongs to the same general class of phenomena as the absorption of various liquids by different fibrous materials and elastic jellies, and it very closely resembles the absorption of water by other cellulosic fibers. Like cotton, wood is a complex fibrous material made up of a number of hollow, elongated cells differing in structure and function and oriented principally in a direction parallel to that of the longitudinal axis of the material. It consists mostly of cellulose, noncellulosic polysaccharides, and lignin arranged in a cellular structure composed mainly of long, hollow fibers typically 1 to 5 mm in length and 20 to

TABLE 10.3

Standard Deviations from the Experiments Comparing Moisture Contents of Beet Pulp by Oven Drying and Rapid Method of Cotton et al.

		Standard deviation		
Experiment no.	Number of Replications	Liquid pressed from pulp (ml)	Moisture by oven method (%)	Moisture by rapid method (%)
1	12	1.0104	0.2291	0.2571
2	10	0.5385	0.0917	0.0755
3	10	0.7000	0.1204	0.0980
4	10	1.2206	0.3470	0.1709
5	10	0.9434	0.1166	0.1356
6	10	1.0198	0.1342	0.1428
7	10	1.1136	0.1600	0.1559
8	4	0.8660	0.1480	0.1212
9	10	1.9209	0.0748	0.2717
10	9	0.4714	0.1723	0.0608
	Average:	0.98046	0.15941	0.14892

40 μm in diameter with cell walls 2- to 8-μm thick oriented in the direction of the tree axis. The fibers are closely packed and cemented together so that the void spaces in wood consist almost entirely of the fiber lumina and the central cavities of other types of cells present (including, in the case of hardwoods, continuously longitudinally oriented cells which may be 200 μm in diameter). By employing X-ray diffraction and electron microscopy techniques, it has been established that the cell wall of the principal structural elements is made up of layers, each layer containing microfibrils oriented in a random manner. Microfibrils about 60 to 200 Å wide, one-half to one-third as thick and indefinitely long are made up essentially of bundles of long-chain cellulose molecules, perhaps 10,000 to 50,000 Å long, which are partially crystalline and partially paracrystalline, i.e., mesomorphous. The paracrystalline regions are

3 MOISTURE IN WOOD AND TIMBER

of principal significance in wood moisture relations, since they provide the sorption sites for attracting and holding water in the cell wall. Water is held in wood both as free and bound water. The bound water is much more tightly held within the cell walls, probably by hydrogen bonding at sorption sites, each of which can hold perhaps six molecules of water.

However, it was recognized early that the sorption rates in most cases of wood samples did not correspond with simple Fickian diffusion of water vapor into the material, and subsequent studies have been concerned with accounting for the departures from ideal behavior. The main lines of evidence that the rate-determining process in thin wood samples is other than diffusion may be summarized as follows:

1. The rate of approach to sorption equilibrium is independent of the thickness of the wood specimen below certain limits, at least throughout the measurable part of the sorption process. Equilibration by diffusion, on the other hand, slows down as the thickness of the specimen increases.

2. For successive equal increments of moisture content over different ranges, e.g., 0 to 3, 3 to 6, 6 to 9% moisture contents, the sorption rate decreases, whereas the diffusion coefficient is known to increase with moisture content.

3. For moisture content increases having the same final value, the time required to reach equilibrium increases markedly the smaller the moisture content change involved. Again, a diffusion coefficient increasing with moisture content should lead to the reverse of this.

An understanding of the interrelationship between wood and moisture and accurate means for control and measurement of such moisture contents are therefore important for the most effective use of wood, timber, and similar materials. The moisture content of wood varies from about 8 to 20% on the dry weight basis. The amount of moisture in wood affects several of its properties. Among these most important is the dimensional stability. Wood shrinks as it dries below the fiber saturation point. Dimensional stability is a requisite of utmost importance in all timber products. Articles made out of timber conditioned to equilibrium moisture content possess maximum dimensional stability. It follows, therefore, that the first step in all timber industries should be to determine the moisture content of the raw material with a view to ascertaining its proximity to the equilibrium moisture content. Besides this, reference to moisture content has also to be made in the commercial transactions of wood, in standardizing its strength properties, and in the various processes involved in wood utilization, as for example seasoning, preservative treatments, woodworking operations, glueing, polishing, etc.

3.2 Fiber Saturation Point

The term moisture content or moisture content percentage of a piece of wood as is commonly understood means the weight of moisture present in timber expressed as the percentage of its dry weight. The word dry indicates the state in which the sample does not have any moisture. Although this definition has been accepted as a standard by the scientific and technical organizations, among wood-based industries and others, there is an obvious incogruity in it. A given volume of wood of different densities at the same moisture content does not contain the same amount of moisture. As a consequence, different timbers have to be dealt with differently with regard to the various processes involved in their utilization. However, there are several important points in favor of this mode of expressing the moisture content of timber. For instance, the shrinkage and swelling of wood, its various strength values, and the logarithm of its electrical resistance are linearly related to its moisture content. By simple instrumentation these properties can be made use of in determining the moisture content of wood. Again, for comparison purposes the adjustment of the strength values of timbers to any standard moisture content can be easily done. Moisture content is also sometimes (for instance in chemical analysis of wood) expressed as a percentage of total original weight of wood instead of its dry weight. The drawback in this definition is that the loss of the same amount of moisture at different stages of the drying does not correspond to the equal changes in moisture content.

3 MOISTURE IN WOOD AND TIMBER

It has long been known that variations in moisture content below a certain limit affect many important physical properties of wood. When it is considered that the strength of wood is greatly affected by absorbed water and not at all by free water, that the swelling and shrinkage of wood are due entirely to absorbed water, and that in drying wood the absorbed water is the more difficult to evaporate, the practical importance of accurate determinations of the limit of absorption, i.e., the fiber saturation point, is at once evident.

The term fiber saturation point was first used in connection with wood to designate the moisture content below which further reduction of moisture causes changes in the strength of the wood. Expressed in the terms that pertain specifically to its cellular structure, the saturation point of a wood fiber may be considered as the state in which the cavity of the fiber is entirely devoid of free water and its cell wall is saturated with bound water throughout. This point is usually obtained at a stage of drying at about 30% moisture content on the dry weight basis. The saturation point marks the level of moisture content below which the moisture changes affect most of the physical and mechanical properties.

3.3 Methods of Moisture Measurement

3.3.1 Oven-Drying Method.
The percentage moisture content of wood or timber is based on the oven-dry weight and is calculated as follows [12-13]:

$$\text{Percentage of moisture} = \frac{W_1 - W_0}{W_0} \times 100 \qquad (10.1)$$

where W_1 is original weight and W_0 is oven-dry weight of the sample. For the determination of moisture content according to the above formula an adequate number of representative sections of wood of suitable size depending on the type of store but not less than three should be taken at random and each accurately weighed. They should be dried in an oven at a temperature of 100 to 105°C till the dry weight of each is constant. Care should be taken to prevent changes in moisture content between the cutting of the section and the first weighing, or between removal from the oven and subsequent weighing. In cases where a sample cannot be obtained in the manner described above, it should be obtained by the use of an auger or bit, boring to a depth of half the thickness of the piece, the borings being collected in a proper receptacle to guard against moisture loss before they are weighed. The borings shall be dried in an oven till a constant dry weight is attained. The oven-dry method as applied to measurement of moisture content of wood and timber is open to the following errors:

1. The final weight varies with the atmospheric humidity. Fluctuations as high as ±0.2% may be obtained.
2. The dry weight attained between 100 to 105°C varies with temperature.
3. Some error is caused by weighing the sample hot from the oven.
4. Sawing of the test sample if done with a dull saw causes appreciable error.

The cumulative effect of all these errors on the determination of the final moisture content generally does not exceed 1%. Besides the errors involved, it has the following drawbacks:

1. The oven-drying method of determining moisture percentage is very slow and elaborate.
2. By this method the moisture content is obtained indirectly. The moisture content of the test section may not be representative of the stock.
3. In cutting the test sections valuable material is lost.
4. The method does not give correct values of the moisture content of wood having volatile oils and resinous matter. If the specimen contains preservative which uses volatile oils as medium such samples cannot be analyzed by oven-drying method.
5. Along with moisture, volatile oils are also distilled off giving considerably high values of moisture content.

3 MOISTURE IN WOOD AND TIMBER

6. Woods go on losing weight even after several days in the oven, making it difficult to fix a correct dry weight of the oven-drying sample.

3.3.2 Distillation Method. The best method for finding the moisture content of wood containing appreciable quantities of substances that volatilize at a temperature below or at the boiling point of water is by distillation. The method consists of boiling the chips of the wood in a liquid immiscible with water and condensing the vapor in a receiver. The water separates out in the condensed product and its volume is measured, from which moisture percent is calculated. Generally, there is an arrangement in the receiver itself for reading the volume. The liquid ordinarily used is toluene. The different distillation methods and techniques of moisture measurement are described in the literature [9, 14], and Chap. III. This method is generally superior to oven drying and usually gives slightly higher results since wood can be completely dried only with difficulty and then must be weighed in a dry atmosphere. Other volatile impurities, such as turpentines, are not included in the results unless they are soluble in water. It is therefore especially valuable for wood chips and sawdust. If only a small sample is available, it is preferable to use the smaller distillation apparatus described in TAPPI Standard T 484 m-58 for paper and paperboard.

The size of the specimen should be such as to yield enough moisture for an accurate measurement, but should not exceed 300 g. The method is accurate to within about 0.2 ml of water. The accuracy of the reported percentage depends on the size of the sample and its moisture content. For a small volume of water that need not be transferred and with accurate calibration of the trap with water under toluene, the accuracy may be increased to be within 0.1 ml. The Dean-Stark distillation apparatus has been modified for the measurement of moisture content of wood chips by Young and James [15].

3.3.3 The Karl Fischer Method. This method has been applied for the measurement of moisture content of wood by Kollman and Hockele

[16]. A detailed description of this method is given in Chap. IV. As is well known, this is a chemical method capable of giving highly accurate results. The procedure involves extracting the water from the fragmented specimen of wood by soaking (a few hours to a few days) in anhydrous methanol, and then determining the water in the methanol after titration with the Karl Fischer reagent. Though the method is capable of giving very accurate results, it is time consuming and requires considerable dexterity on the part of the operator for proper experimentation. Also the solutions require frequent standardization and have to be kept completely free from atmospheric moisture as they are highly hygroscopic. However, during the last 20 years the Karl Fischer method for estimating water chemically has been considerably extended in scope and improved in instrumentation and is now a standard technique in most laboratories. Alternative solvents and new methods of endpoint detection have increased the accuracy of the method and have made it possible to exploit the versatility of the Karl Fischer reaction to a high degree.

3.3.4 Electrical Method. The oven-drying, distillation, and Karl Fischer methods of moisture determination of wood are rather difficult to adopt in commercial practice as they are time consuming and destructive and involve complicated procedures. An arrangement which is portable, robust, easy to operate, capable of giving direct and quick readings preferably in situ, and reasonably accurate is required. Electrical moisture meters which first appeared in the early 1930s satisfy most of the conditions and are now being successfully employed in wood-based industries for commercial measurement and control of moisture in the processing of timber and in the manufacture of wood products. They have made significant contributions towards raising the standard of timber industries. Electrical moisture meters also give satisfactory service in testing resinous timbers as well. Two types of electrical moisture meters are available. These are the conductivity and radiofrequency power-loss types.

3 MOISTURE IN WOOD AND TIMBER 677

A compact and portable electrical conductivity or resistance-measuring apparatus (Fig. 10.6) designed by Stamm [17] is quite satisfactory for determining the moisture content of wood. It consists of a portable Leeds and Northrup suspension galvanometer G, an inexpensive student-type Ayrton shunt S, four small 22.5-V batteries B connected in series to supply the potential of 90 V, a 3-V flash lamp battery F for lighting the galvanometer lamp, a toggle switch T to operate the galvanometer lamp, and a momentary contact push-button switch K for closing the galvanometer circuit. The galvanometer is connected across the shunt as indicated in the figure, and the shunt in turn is connected in series with the 90-V battery, the pin contacts, and the push-button switch. The needle electrode system N is connected to the galvanometer G through the battery B as shown in Fig. 10.6.

From the moisture content electrical conductivity data a galvanometer scale is prepared to read directly in terms of moisture content. Four different moisture content scales, corresponding to the four different shunt settings of 1.0, 0.1, 0.01, and 0.001, respectively, enhance the range of the instrument. Each of the scales is in a different color and the corresponding shunt settings are

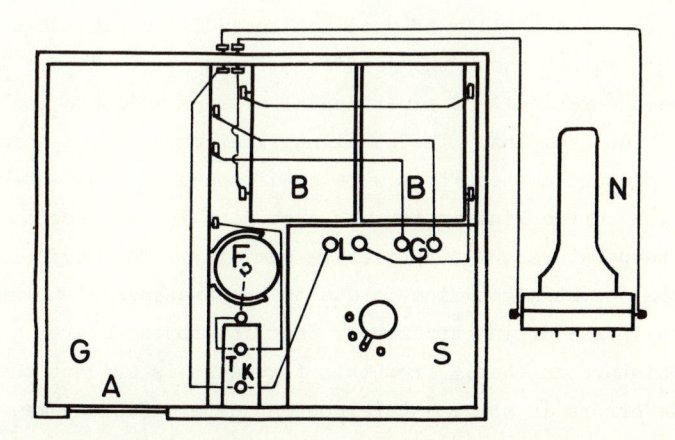

FIG. 10.6. Schematic diagram of complete moisture meter for wood.

painted in the same color to simplify the reading of the instrument. The glass A carrying the galvanometer scales is made movable to facilitate adjustment to the correct zero setting. A resistance of 1000 ohms is connected in series with the galvanometer inside the galvanometer case for protection. This resistance prevents short-circuiting through the galvanometer circuit.

Electrical conductivity measurements, using the pin-type contact, were made on blocks 7 x 5 x 2 cm. Stamm [17] examined 25 different kinds of wood samples. These woods included many, if not most, of the commercially important softwoods and hardwoods. The density range was from 0.346 to 0.705 g/cm^3. Some of the woods were high in extractives and resin content; others were low. The electrical conductivities of these specimens were plotted [17] logarithmically (Fig. 10.7) against the average moisture content values as determined from the loss of weight upon oven drying at 105°C. An approximately straight-line relationship was obtained. A tendency for the straight line to curve to the right may be expected for moisture content values that approach the fiber saturation point.

The data shown in Fig. 10.8 which were obtained by Stamm [17] by using surface contact electrodes for the measurement of the conductance of wood specimens illustrate very well the effect of moisture gradients in a piece of wood and in addition give some interesting information about the gradients themselves. Figure 10.9 shows the theoretical and experimental average moisture content values of wood samples. The straight line represents specimens of uniform moisture content that were conditioned in constant-humidity rooms. The curved line is for wet specimens in the process of drying at room temperature and 30% relative humidity. The deviation of the curve from the straight line is due to the presence of moisture gradients in the drying specimens, that is, to the uneven distribution of the moisture in them. From this figure it is evident that considerable errors in the moisture content determinations may result from using the surface-contact type of electrodes when appreciable moisture gradients are present.

3 MOISTURE IN WOOD AND TIMBER

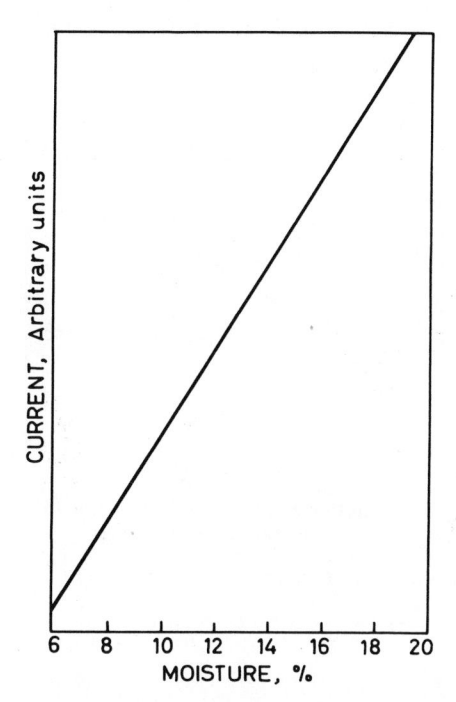

FIG. 10.7. Graph showing relationship between logarithm of electrical conductivity and moisture content.

It was observed by Stamm [17] that southern yellow pine, which is very high in resin content, gave moisture content readings from the electrical measurements that were several percent lower than the values from oven drying. It was suspected that the oven-drying results were too high because of loss of resin as well as loss of water in drying. The moisture content, therefore, was determined by dry distillation at 105°C with condensation of the distillate. The values of moisture content as determined from the volume of the water distillate agreed very well with the electrical conductivity values. This agreement tends to show that the moisture content as determined by the electrical conductivity method may be more accurate for resinous woods than the values obtained by the change in weight on oven drying.

680   X. MOISTURE IN BAGASSE, WOOD, AND PAPER

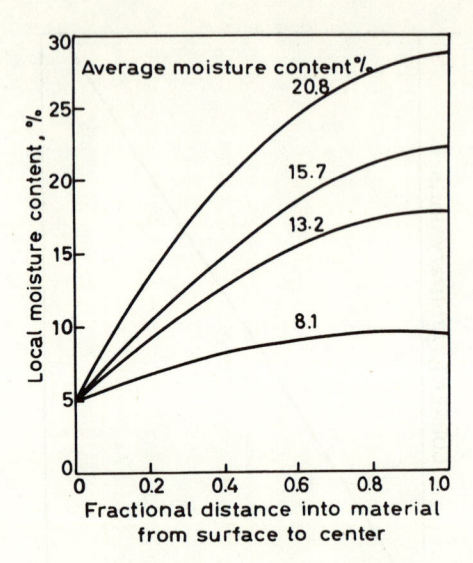

FIG. 10.8. Graph showing the moisture gradients at different depths in wood.

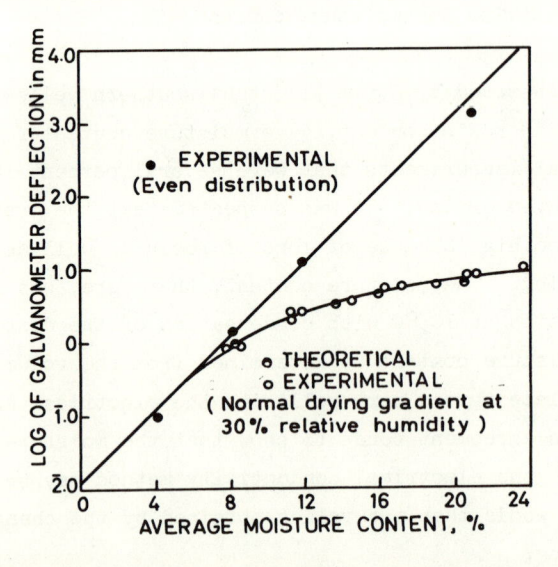

FIG. 10.9. Theoretical and experimental average moisture contents of wood.

3   MOISTURE IN WOOD AND TIMBER   681

The electric moisture meter though very valuable for control purposes cannot be used for finding the absolute moisture content accurately without proper calibration. Below are mentioned some of the factors which affect the accuracy of the resistance moisture meter:

1. The resistance moisture meter is not a very sensitive instrument. Generally the least reading corresponds to a moisture content difference of 1%. It has limitations in the measurement of very low and very high moisture contents.

2. Specimens of the same species at the same moisture content may not give the same reading in the moisture meter because of the difference in their extractive content or density. Error due to these causes is generally small.

3. The electric resistance of different species of wood at the same moisture content is generally not the same. Naturally, the moisture meter will indicate different readings for these species. This disadvantage is overcome by preparing correction tables for the readings of the moisture meter for different species.

4. When the surface of the timber is better than the interior, the reading of the moisture meter will record a much higher moisture content than the average. Moisture pockets in the interior are also the cause of higher readings. They can, however, be detected if moisture content is taken at more than one position.

5. Temperature change is another source of error.

6. In wet weather, the surface insulation of the moisture meter becomes wet and tends to leak, thus seriously interfering with the accuracy of the instrument.

7. Timbers which have been immersed in water and treated with water-soluble salts or surface coated do not give reliable results. The same is the case with the material hot from the kiln.

8. Error in the reading of the moisture meter can also be due to the thickness effect, unequal moisture distribution, or the electrodes not being inserted along the grain.

In the 7 to 25% moisture range, the accuracy of the resistance-type instrument when properly calibrated and correctly used should be within ±1%. It is not expected that readings of moisture content above 25% will be as accurate as those in a lower range, nor do they ordinarily need to be. The readings of the moisture meter do not hold good at temperatures other than those for which they have been calibrated. For observations on temperatures other than the temperature of calibration, use of a temperature correction chart (Fig. 10.10) should be made. This figure shows the temperature corrections applicable to moisture contents as determined with resistance-type electric moisture meters.

Correction tables have been prepared by Rehman et al. [18] for the readings of a number of commercially available resistance or conductivity moisture meters with reference to some commercially important Indian timbers. A special plate electrode has also been developed by Rehman et al. [18] for veneers and other thin stock. It has two or more concentric coplanar metallic rings which serve as electrodes. The readings obtained with it are correct only for materials having uniform moisture content.

The dielectric loss factors of wood and paper depend on their moisture contents. The power-loss moisture meter measures the loading of a radiofrequency circuit when the plates of a capacitor coupled to the circuit are pressed against the wood and paper samples, whose moisture content is determined from the correlation of power-loss factor as shown in Fig. 10.11.

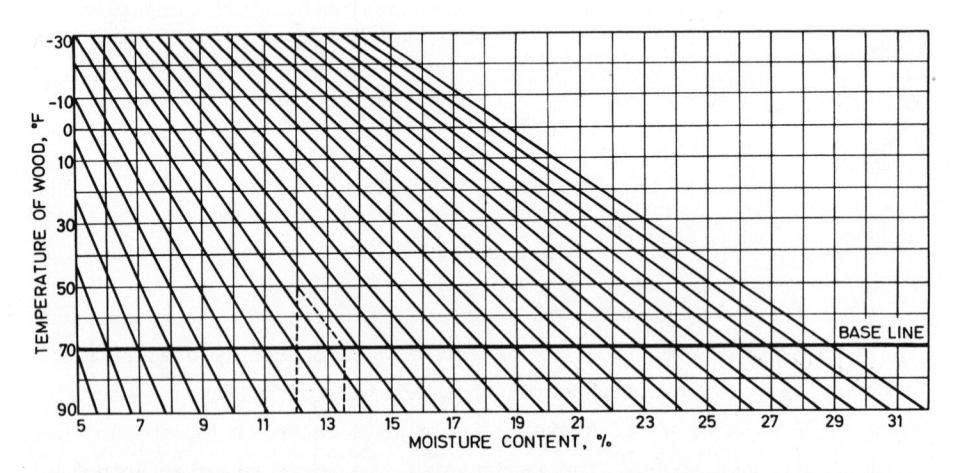

FIG. 10.10. Graph showing the temperature corrections (below 90°F) for resistance-type moisture meters for wood.

3   MOISTURE IN WOOD AND TIMBER   683

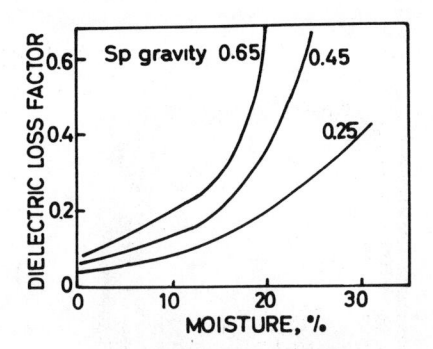

FIG. 10.11. Relationship between dielectric loss factor and moisture content of wood at various levels of specific gravity.

The instrument known as the Moisture Register (Fig. 10.12) consists of two units connected by a flexible rubber-covered cable. The contactor, or sample holder, which is pressed against the sample is mounted on the end of the first unit, called the gun, in which are contained a radiofrequency oscillator tube and its associated circuit components, a spring arrangement for applying pressure to the sample and a microswitch which closes when the proper pressure is applied. Power for the gun is supplied through the cable from the battery and meter unit. Within this latter unit are located plate and filament batteries for the oscillator, a push-button switch for these batteries, a small lamp which lights up when the microswitch is closed, a micrometer, a potentiometer for zeroing the micrometer, and various other resistors.

The operation of the instrument is as follows: The gun is held away from any absorbing materials, and a push button is held in, which closes the two battery switches. After the oscillator has warmed up, the potentiometer is adjusted until the micrometer reads zero (full-scale deflection). The gun is then pressed against the surface of the sample of wood or paper whose moisture content is to be determined, the pressure being increased until the lamp lights, and a reading is taken on the micrometer. From the calibration curve, the micrometer reading is transformed to a moisture content value.

684   X. MOISTURE IN BAGASSE, WOOD, AND PAPER

FIG. 10.12. Photograph of Moisture Register.

   This instrument is connected in the circuit so that it measures
the current drawn by the grid of the oscillator tube. This grid
current depends on the amplitude of the oscillation which, in turn,
depends on the losses of the tuned circuit. The contactor is coupled
to this tuned circuit and acts as a condenser, a part of whose dielectric is the paper. Dielectric losses in wood and paper decrease
the amplitude of the oscillation and the magnitude of the grid current. As these losses increase with the moisture content of the
material the grid current becomes a measure of the moisture content.
The grid current is also affected by the plate and grid voltages,
and the potentiometer which adjusts these voltages is a zero adjustment, which is used to correct the changes in battery voltages.

The operating frequency of the Moisture Register is 10.4 MHz when there is no sample against the contactor. If the block of wood or pad of paper whose moisture content is to be measured is such that the grid current drops to about half its original value, the frequency is reduced to about 8.4 MHz. If the grid current is decreased to zero, the circuit stops oscillating. The extent of the field about the contactor is limited by a pair of guard rings around the electrodes, so that it penetrates only a small distance into the paper. The instrument is usually calibrated with a sample of wood and paper thick enough so that additional sheets do not change the reading. For this purpose, a pad 3 mm in thickness is recommended by the manufacturer.

Unfortunately, no material is available whose dielectric loss can be standardized. The Moisture Register is therefore forced to rely on the stability of the circuit itself to maintain the shape of the calibration curve. Most of the drift which normally occurs in a vacuum tube circuit is balanced out, however, by means of a zero adjustment, the meter being brought to full-scale deflection (zero reading) with the contactor held away from the sample.

A new machine has recently been developed by James [19] that is used for measuring the moisture content of moving lumber, veneer, or paper. In particular, it is designed to mark or reject automatically material with a moisture content improper for use. It is used for sorting lumber or veneer on the dry chain, or inspecting material approaching cutting tables or planers. This machine is a combination of the resistance-type and capacitance-type meters.

4 MOISTURE IN WOOD CHIPS

4.1 Introduction

In the manufacture of paper and paper products from wood pulp, the need for proper and accurate control instrumentation is as imperative as in any other industry. Throughout the entire process, information is required that will enable the operators to operate along the thin line that separates economy and inefficiency. Such control instrumentation is necessary at every point from the wood yard to the storage of the product. One outstanding example of such a point is the wood chip conveyor line. For many years the continuous and dynamic weighing of the wood chips has been an accepted fact. This information has been used to control the digester operation at the most efficient and practical level. However, this information is not nearly as significant as it could be if the moisture content of the chips were also known. Where the moisture level varies considerably, the need is even greater than in those areas where it is relatively constant. In either case, it must be monitored to allow correction for variation when it does occur. The actual monitoring of a stream of flowing wood chips is rather a complicated affair.

Perhaps the most difficult aspect to handle is the sampling problem. Although it is practical to sample on a batch basis for a laboratory instrument, this is out of the question for a process device such as in this application. Not only must the instrument be capable of sampling the continuous flow of wood chips, but also it must do so without seriously affecting the passage of the material. To accomplish this, the wood chips must pass by the surface of the measuring device and be scanned or analyzed as they pass.

4.2 Moisture Measurement

There are a number of methods available for this purpose. These are as follows:

1. Conductivity method
2. Capacitance method
3. Differential temperature method
4. rf Power absorption method
5. Ambient relative humidity method
6. Radioactivity (neutron absorption) method

In the case of the conductivity method the main disadvantage is the surface phenomena generally encountered. Since the conductivity of a material is a function of the surface contacted by the measuring electrodes, that moisture which is inside the solid piece under consideration is not observed by the instrument. In the case of wood chips this is particularly true since their immediate history before measurement will materially affect the accuracy of the reading. As a second objection, the measurement of the conductivity of wood chips would be extremely difficult due to their rough surface and unpredictable shape. Measurement of temperature difference is also not applicable here for the same reason of heterogeneity, as well as the fact that in the environment where it is necessary to measure the moisture content of the wood chips the ambient conditions vary over a wide range, thereby introducing errors.

The use of radiofrequency power absorption in the case of wood chips is impractical for many reasons. First, the radiofrequency field intensity must be prohibitively large to ensure representative sampling. This increases the effect of stray fields. In addition, the sample handling problem is almost insurmountable because the weight or volume of the material within the radiofrequency field must be known exactly. Therefore, although this approach is used with success in laboratory applications, it is not practical for a process instrument in the case of wood chips. Of all the analytical methods, the measurement of the relative humidity immediately adjacent to the material as a function of its moisture content is perhaps the least attractive. First, the rate of evaporation of water from the material being measured is not constant. Second, the instrument accuracy is affected radically by changing moisture content in the air surrounding the measurement area. The measurement of moisture content by radioactive emission is relatively new in the field of instrumentation. It was initially developed for the measurement of soil moisture in construction work. It is still largely used for this purpose with great success. Its application to the measurement of moisture in wood chips has certain drawbacks.

To overcome the disadvantages attributable to the foregoing methods, the capacitance or dielectric system of moisture measurement has been developed by Shaffer [20]. The high dielectric constant values of water as compared to that of wood ensures ease of distinction between the parameter to be detected, in this case water, and the parent material, wood. This is true because a relatively small variation in moisture content of moist wood causes appreciable change in its dielectric constant. Shaffer's apparatus consists of an oscillator, the sensor, condenser being part of the capacitance forming the resonant circuit of the oscillator. When the dielectric constant of the material within the field of the sensor changes, its capacitance changes. Such a change causes a corresponding shift in oscillator frequency. This oscillator is, therefore, followed by a conventional discriminator network, which produces a voltage as the oscillator frequency is shifted from the design center. This voltage produced at the discriminator is interpreted, through calibration, as water content.

The automatic frequency control system was adopted for controlling the frequency of the oscillator. This system repetitively compared the sample signal to a temperature-compensated "standard" capacitor. The resulting frequency modulation of the oscillator output was sensed by a conventional automatic frequency control circuit and used to correct the oscillator-tuned circuit so that the average frequency was held at the center of the discriminator response curve. In order to obtain a dc signal proportional to the moisture content of the sample, the discriminator output was synchronously demodulated. With this system, and again without temperature compensation drift due to temperature change, aging of components was almost completely eliminated. The stability of the oscillator is shown in Fig. 10.13. This run was made after a 6-hr warm-up period. Using this automatic frequency control system the only significant error was in the sensing capacitor. Investigation revealed that erroneous capacitance effect from the sensor could be attributed to three main areas. The first error was produced by thermally induced physical change in the capacitor itself. Since application demands a strong construction, many thermal changes can occur, often with long equilibrium delays. Temperature compensation was therefore effected [20] through the use of negative temperature coefficient trimmer capacitors placed in the same thermal environment as the sensor. Stability achieved by Shaffer was equal to 0.07% drift for a 100°F change. The complete instrument developed by Shaffer is shown in Fig. 10.14.

5 MOISTURE IN PAPER AND PAPER PRODUCTS

5.1 General

The problems in the measurement and control of moisture in paper are very similar to those in textiles as both are fibrous materials and the moisture is absorbed or desorbed according to the same laws of diffusion and capillary action which have been described in the case of textile materials in Chapter IX. The moisture content of paper is a quantity of considerable interest both to the manufacturer and to the user. The physical properties of paper such

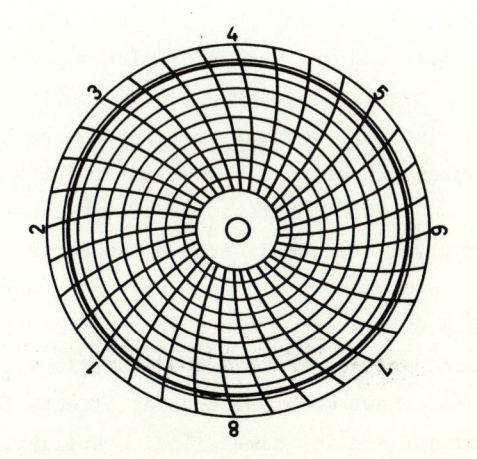

FIG. 10.13. Chart record showing stability of the reading after using automatic frequency control.

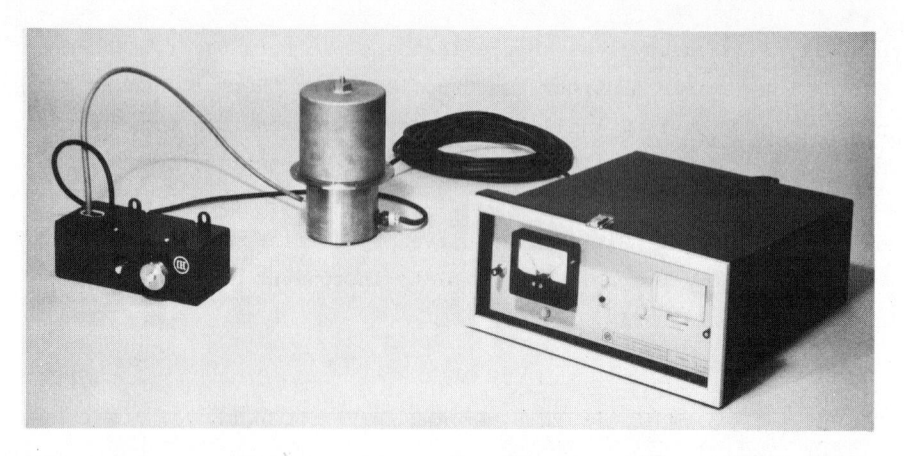

FIG. 10.14. Complete instrument (dielectric type) for continuous measurement of moisture content of wood chips.

as strength, basis weight, tendency to curl, electrical resistance, and dielectric properties are profoundly affected by its moisture content. A change in the moisture content causes a change in the dimensions of the sheet, causing constraint on the printer. Every paper technologist knows that in paper-making, coating of the sheet or board is very important. The quality of the finished product depends to a great extent on the amount of moisture present at critical points in the process. Overwet paper may crush or mottle in the calender rolls and the desired finish may never be obtained. On the other hand, overdried paper, in addition to suffering from the standpoint of quality, becomes costly because it loses in weight. Every pound of water unnecessarily removed from the sheet diminishes the net profit. In coating operations, if the paper is too dry when the coating is applied, the bond between the paper and the coating is often unsatisfactory and the coating may pick on the printing press. Conversely, too much moisture is often responsible for the coating lying improperly on the surface. Similarly, in many converting operations, sheet moisture content affects the quality of the finished product, and its dimensional stability.

As is well known, dimensional instability is caused in paper during the process of absorption and drying of paper in the same

5 MOISTURE IN PAPER AND PAPER PRODUCTS

manner in which it occurs in textile materials. Before it is generally possible to understand the reasons for the existence of dimensional instability and other defects in paper which are caused by variations in moisture content, it is necessary to have an understanding of the structure and composition of paper. Basically, most papers are composed of cellulose (vegetable) fibers, whose most important property lies in their natural ability to swell in water. It is this feature that is made use of during the stages of stock preparation, when the absorption of water molecules in the cellulosic material of the fibers takes place to produce certain physical changes in its structure. The degree to which such water absorption is allowed to take place is carefully controlled according to the requirements and physical characteristics of the finished paper. Therefore the amount of treatment imparted to the fibers will determine the extent to which fiber-to-fiber bonds will be formed in the paper during drying on the paper machine. Thus, the process of drying is of prime importance in relation to the formation of strength bonds and in respect to the study of the problem of dimensional stability and similar properties.

Many of the phenomena that are associated with dimensional instability are related, such as curling, waving, and cockling, and, apart from incorrect handling after manufacture and conditioning, may largely be attributed to irregular substance control and drying rate on the paper machine. For optimal strength and ease of handling in printing or conversion processes, paper and board should contain 5 to 8% moisture in the finished sheet. A longitudinal streak of 12% moisture content in a sheet with 8% moisture overall causes serious wrinkling in the roll. The overdrying needed to correct this represents a double loss, additional steam is consumed per pound of paper produced, and it is necessary to slow down the machine to allow for the additional drying time which results in lost production time.

5.2 Methods of Measurement

5.2.1 Oven-Drying Method. The "paper maker" refers to the moisture content of the moist weight of the paper, whereas "spinning master" refers to the moisture content on the dry weight basis. For mathematical calculations to be made in order to convert one set of moisture content values to the other, the reader is referred to an excellent publication by Brecht [21]. The values of moisture content as obtained on the dry weight basis as compared to the same on the wet weight basis are graphically represented in Fig. 10.15. It is quite convenient to convert the moisture content values from dry weight to wet weight basis and vice versa using the graph given in Fig. 10.15.

In the manufacture of paper and paper products, the value of the moisture content varies from almost 100% to about 7%. A profile of moisture content variation in the process of the manufacture of paper is shown in Fig. 10.16. In principle, all the methods which have been described in Chap. IX for the measurement of moisture

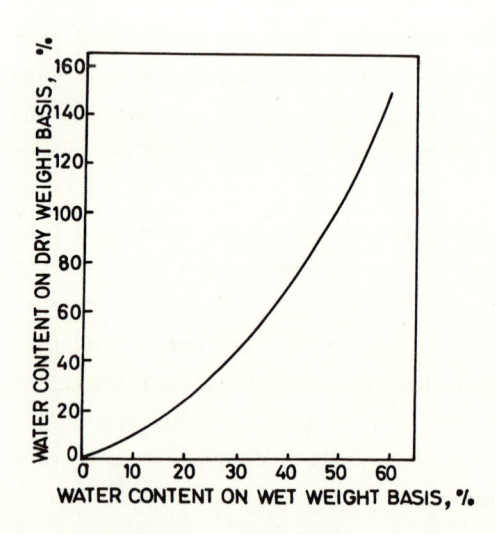

FIG. 10.15. Graph showing the relationship between the values of moisture contents obtained on dry and wet weight bases.

5 MOISTURE IN PAPER AND PAPER PRODUCTS   693

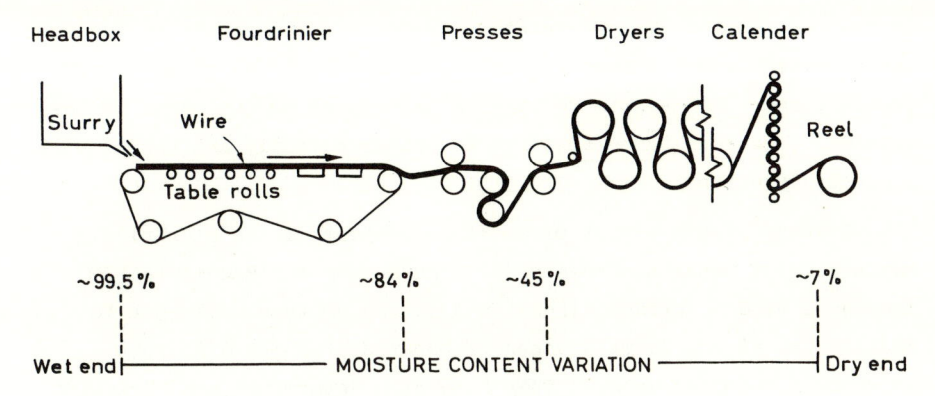

FIG. 10.16. Moisture-content profile from wet end to dry end in the manufacture of paper.

content of textile materials can be applied to paper as well. In the oven-drying method, the procedure followed is that given in the ASTM and other standards. This method, known as ASTM D 644-55, covers the procedure for determining moisture in all paper, paper boards, and paperboard and fiberboard containers, except those containing matter other than water that is volatile at 105°C. The moisture content as the percentage loss in weight of the specimen to the nearest 0.1% on the basis of the original weight or on the oven-drying basis should be reported. The results of duplicate determinations of moisture should agree within ±0.2%. Because of the extremely hygroscopic nature of cellulose, the dried sheet must be enclosed to prevent absorption of moisture while weighing. As the fibers tend to decompose, the drying should not be continued too long. Other precautions described in Chap. II should be observed to eliminate sources of error inherent in an oven-drying method.

5.2.2 Conductivity or Resistance Method. The effect of moisture on the electrical resistance of paper depends a great deal on the mechanism by which the water molecules are bound to the fibers. At low relative humidities, the sorption seems to be a surface phenomenon, whereas for higher humidities, additional moisture is held by capillary forces. The transition between these two also appears to occur in the neighborhood of 50% relative humidity. The resistance changes are affected by the phenomenon of dc polarization and these changes become larger and occur more rapidly as voltage on the paper is increased. The resistance at first decreases slightly and then increases appreciably.

A study of the effect of moisture content on the electrical resistance of paper has been made by Hardacker and Rawcliffe [22]. They have used ac bridge circuit and dc vacuum tube voltmeter to measure the ac and dc resistance of paper containing different amounts of absorbed water. They have also determined the variation of the resistance with time. They have discussed the accuracy as well as the limitations of the moisture determination based on the resistance properties by evaluating two commercial moisture meters working on the principle of the variation of the ac or dc resistances at different humidities or moisture content values. They have concluded that in order to make these moisture meters applicable to paper it is necessary to have separate calibrations made taking into account the variations of the resistance moisture characteristics due to variations in fiber finish, beating, dyes, size, density, and ash content of the samples.

If the paper moves, the relations between the resistance values and moisture content of paper are different. The conductivity will be overshadowed by charging currents of geometrical and internal capacitances. The paper movement can be considered as replacing the dc current by an ac current of a proper frequency. For example, a paper speed of 1500 cpm with 1-in.-wide electrodes would correspond to a frequency of 600 Hz. The apparent resistance will represent the total dielectric losses in the material as well. Figure 10.17 shows the variation of apparent resistance calculated [22] from dissipation factor measurements at different frequencies. It shows that a resistance measurement on a stationary sample cannot be transferred to a moving one.

Based on the resistance variations of a paper sample in accordance with the variation of the moisture content, a number of moisture

5 MOISTURE IN PAPER AND PAPER PRODUCTS

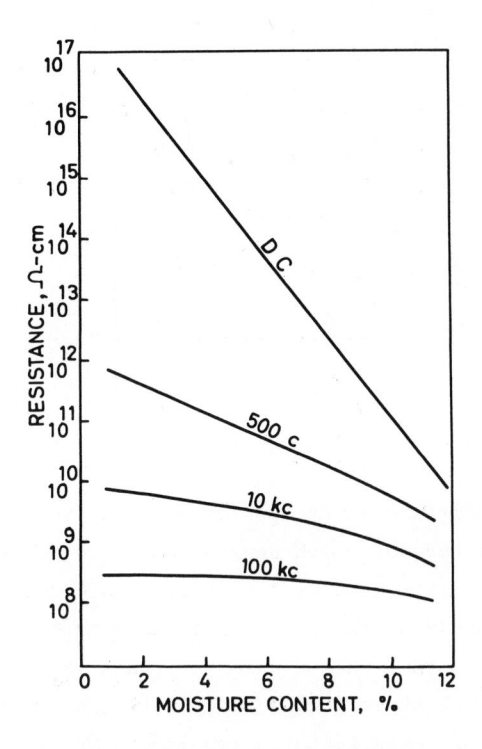

FIG. 10.17. Graph showing the variation of apparent resistance calculated from dissipation factor from measurements at different frequencies.

meters have been commercially developed. Two of these, viz., the Hart moisture meter (Fig. 10.18) and the Moisture Register (Fig. 10.19) have been used for the measurement of moisture content of paper having different amounts of moisture contents at different relative humidities and ambient temperatures. Also a dc vacuum tube voltmeter and ac bridge method have been developed [23] for the measurement of the moisture content of paper as well as other paper products. The design features of these instruments are described in detail elsewhere [22]; here only their applications to the measurement of moisture content of paper are discussed.

Comparison of probable errors of resistance measurements on paper using dc vacuum voltmeter, ac bridge, and Hart meter is given in Table 10.4.

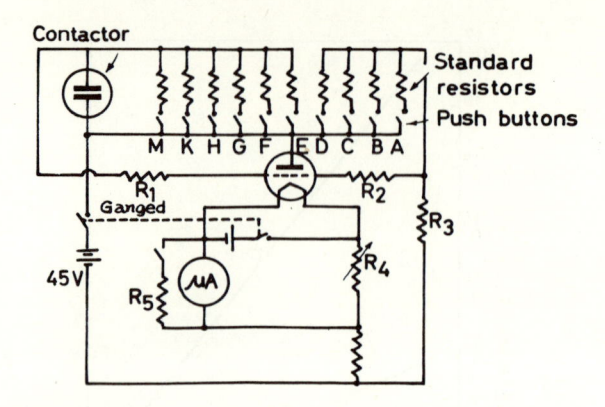

FIG. 10.18. Electrical circuit of Hart moisture meter.

The Hart moisture meter applies a direct voltage across the paper, the magnitude of which depends upon the resistance of the paper and varies from about 1 V to nearly 45 V. It is well known that the resistance of the paper sheet depends on the voltage applied, and also that when a direct voltage is used, polarization effects within the paper become important. In order to determine the importance of these factors, two laboratory instruments were constructed by Hardacker and Rawcliffe [22], one a direct current vacuum tube voltmeter and the other an alternating current bridge, in both of which the voltage applied across the paper could be kept constant. For each of these, the contactor from the Hart meter was utilized to firmly hold the paper sample in order to standardize the clamping conditions.

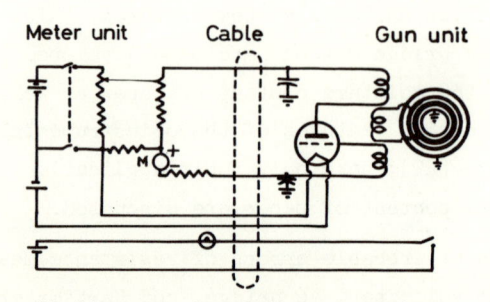

FIG. 10.19. Circuit diagram of moisture register.

5 MOISTURE IN PAPER AND PAPER PRODUCTS

TABLE 10.4

Comparison of Probable Errors of Resistance Measurement of Paper by Hart Moisture Meter, ac Bridge, and dc Vacuum Tube Voltmeter

Sample no.	RH (%)	Probable error (%)		
		ac Bridge	dc VTVM	Hart moisture meter
1	32	0.6	2.1	2.2
2	43	2.9	3.8	3.6
3	48	1.0	1.1	1.7
4	53	1.4	1.0	1.8
5	59	1.9	2.9	0.8
6	68	2.0	6.6	4.0
7	75	5.3	7.0	0.7

Another commercial moisture meter known as Aquatel has been developed by Electronic Automation System (US). The Aquatel system measures ac impedance, but applies electrodes to only one side of the sheet. This avoids the effects of sheet thickness or basis weight that are encountered in devices where the current passes from one side of the sheet to the other. The impedance seen by the Aquatel detector electrodes includes three major elements: sheet resistance, sheet capacitance, and coupling capacitance between the electrodes and the sheet. The measuring circuit (Fig. 10.20) establishes null balance for the resistance and capacitive elements separately. The null-balance principle minimizes errors due to energy losses in the reactive circuit and temperature variations. A logarithmic wound potentiometer in the resistance balancing circuit is mechanically coupled to a linear potentiometer in the output circuit to give a reading which is essentially linear with percent moisture. The Aquatel has ranges of 5% of moisture with a precision of 0.5%/range (0.025% of moisture content). Repeatability of 0.25% for paper and paperboard, 4 to 11% for newsprint can be obtained. Arrangements have been provided for compensating for changes in the

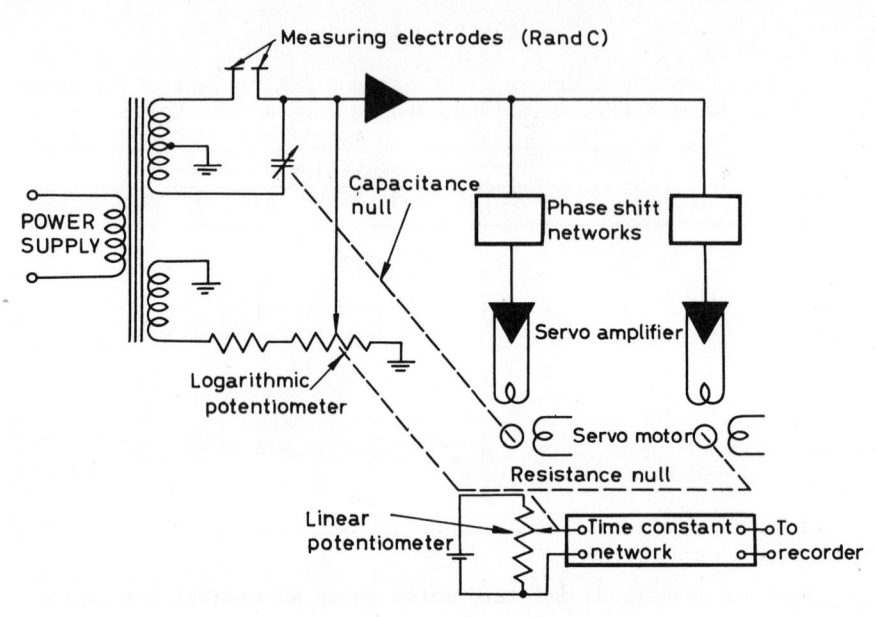

FIG. 10.20. Aquatel ac impedance measuring circuit.

slope of the moisture-conductivity relationship with various compositions of fibrous and nonfibrous materials.

5.2.3 *Dielectric or Capacitance Method.* It is generally known that dielectric properties of a hygroscopic web are functions of relative moisture content, temperature, and composition of the sample. In addition to this the measurement techniques require knowledge of the weight of the sample. Assuming that the composition of material is constant and the equipment is error free, it follows that in general any dielectric measurement to determine moisture content has to be accompanied by simultaneous measurements of basis weight and temperature. The changes of dielectric properties with changes in frequency are explained by changes in polarization. With increase of frequency, different types of polarization effects decrease rapidly in the different frequency regions. Variation of dielectric properties of paper in relation to moisture content has been utilized for developing dielectric- or capacitance-type moisture meters.

5 MOISTURE IN PAPER AND PAPER PRODUCTS

The paper under test or control (at rest or moving) is made the dielectric of an electrical capacitor. In its simplest form, the capacitor consists of a pair of plane, parallel metal plates separated by a distance d. The paper of thickness t only partially fills the space between the plates, so that the capacitance of unit area of the capacitor is given by

$$C = \left(\frac{1.11}{4\pi}\right)\left[d - t\left(1 - \frac{1}{e_m}\right)\right] \tag{10.2}$$

The quantity e_m is the dielectric constant of the paper. If the paper completely filled the space between the metal plates, so that t would become equal to d, the capacitance would increase in value by $1.11 e_m/4\pi d$ pF/unit area. If only air is present between the plates, the capacitance per unit area is $1.11/4\pi d$.

The contribution of a sheet component to the observed dielectric constant does not bear a linear relation to the mass per unit volume of that component. It was shown by Van den Akker and Hardacker [24] that experimental data on the dielectric constant of dry fiberboard covering a density range from insulation-type board to hardboard showed that the Clausius-Mosotti relation fitted the experimental observation with good accuracy. Delevanti and Hansen [25] have shown that this function may be similarly employed for handling data on the dielectric constant and loss angle of paper sheets; the function has been similarly employed by Calkins [26] and recently Verseput [23] has shown that it fits with good accuracy data for sheets prepared from regenerated fibers. In the form employed, the relation is

$$\frac{e - 1}{e + 2} = K\rho \tag{10.3}$$

in which K is a proportionality constant, ρ is the density of the sheet, and e is its dielectric constant.

It has been estimated that the change in dielectric constant caused by a moisture content change of 1% is 0.049, a rather small quantity. Thus the expectation of the change in dielectric constant of typical paper corresponding to a 1% change in the moisture content at 5% is not nearly as great as many persons have supposed on the implied assumption of approximate linearity between the density and the contribution to the dielectric constant. This change of 0.049 is about 2% of the dielectric constant of the paper and water mixture. The smallness of the change in the dielectric constant should not, however, deter one from using the capacitance method for the measurement of moisture content. Another factor to be taken into consideration relates to the properties of the paper itself, whose moisture content is to be determined. The distance between the two plates of the electrode system forming a condenser in which the sample is placed for the moisture measurement is very critical and the sensitivity of the moisture meter is very dependent on this spacing. Calculations with Eq. (10.3) show that an air space between the dielectric material and the electrodes has a strong desensitizing effect. Lastly, the most important consideration is by what electrical method very small change in capacitance should be measured accurately. There are many possibilities and a number of instruments have been developed which have been successful to varying degrees in measuring the moisture contents of paper and paper products.

Fielden [27] has utilized the principle of the alternating current capacitance bridge method and developed a moisture meter which is known as the Drimeter. The Fielden Drimeter, although designed primarily for the control of moisture in textiles, has been reported to serve with some success in the control of moisture content of paper products in England.

As a result of the investigations carried out by Van den Akker and Hardacker [24] it was concluded by them that the principle of controlling moisture content through its relationship with the dielectric constant of a material is sound in theory. The Drimeter, however, was judged to be unsatisfactory for the measurement and control of moisture content of paper and paper products.

Thode [28] has developed a dielectric moisture meter based on the change of capacitance or admittance of the moist paper. This meter works efficiently even on low frequencies and high moisture content values of the order of fiber saturation point. This has been made possible by the development of a fringe field probe which overcomes the difficulties encountered with the use of parallel plate condensers having very narrow gaps between their elements. The fringe field probe (Fig. 10.21) consists of an array of condenser electrodes embedded in a good dielectric with thin guard electrodes sandwiched in between. The polarity and potential of the measuring electrodes and the guard electrodes are arranged so that within the dielectric, no field exists between the measuring electrodes. All capacitance between measuring electrodes is in the air gap at the surface of the probe. The fringe field probe developed by Foxboro Company has a capacitance of 20 pF in the air gap, with no paper present. Five parallel strips form the active electrodes, three of one polarity and two of the other. The electrodes are arrayed across the paper machine transverse to the sheet's direction of travel.

Not all commercial devices of this type depend on capacitance measurement alone. The Industrial Nucleonics moisture gauge, Fig. 10.22, determines total admittance at two frequencies simultaneously. These two values, plus a temperature measurement from a thermistor element, feed an analog computing circuit that produces a temperature-corrected output. This output is said to be independent of basis

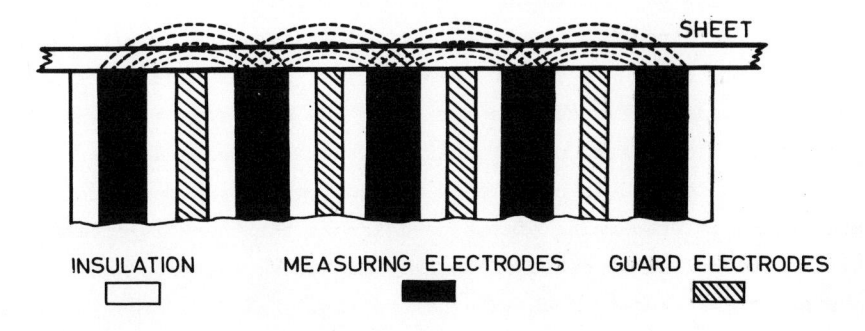

FIG. 10.21. Fringe field probe.

702   X. MOISTURE IN BAGASSE, WOOD, AND PAPER

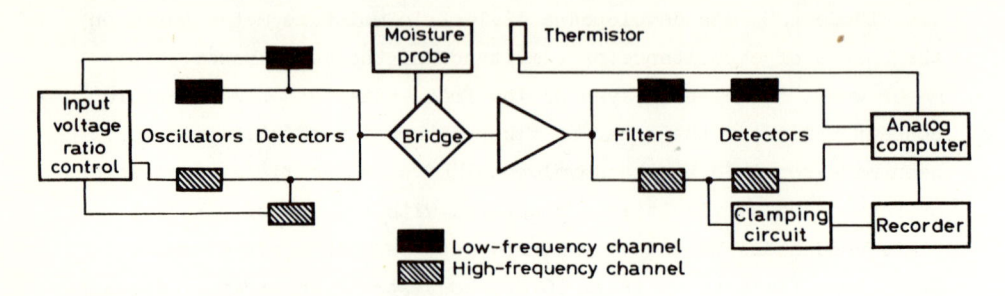

FIG. 10.22. Nucleonics moisture gauge.

weight and quality of the probe contact. The detector head is a fringe field capacitor probe with a large surface containing many strips running in both the machine directions. The designers claim that this electrode configuration eliminates the effect of fiber orientation. The probe connects to a bridge circuit that is excited by two different oscillators operating at widely separated frequencies. After linear amplification, the two signals from the bridge unbalance are filtered, further amplified, and rectified. From the resulting dc signals, E_1 and E_2, a simple computing circuit produces a single output voltage E according to the following relation:

$$E = (E_1 - E_2)E_1 \qquad (10.4)$$

The Honeywell Aquaset is another capacitance-measuring device employing a single-sided contactor that contains two measuring capacitors and two reference capacitors in bridge circuit (Fig. 10.23). The circuit compensates for temperature and humidity. This 5-MHz system has an automatic standardizing circuit and adjustment potentiometers on the console for range and zero and basis weight. The head, which rests on the paper sheet, is raised and lowered by an air cylinder and is capable of various swings and positioning adjustments. Traversing of the head across the sheet is accomplished by a motor driven by cable-and-pulley mechanism supported on a channel beam. For a cross-machine profile, an X-Y recorder accepts the transverse position of the head as one input, and the moisture

5 MOISTURE IN PAPER AND PAPER PRODUCTS   703

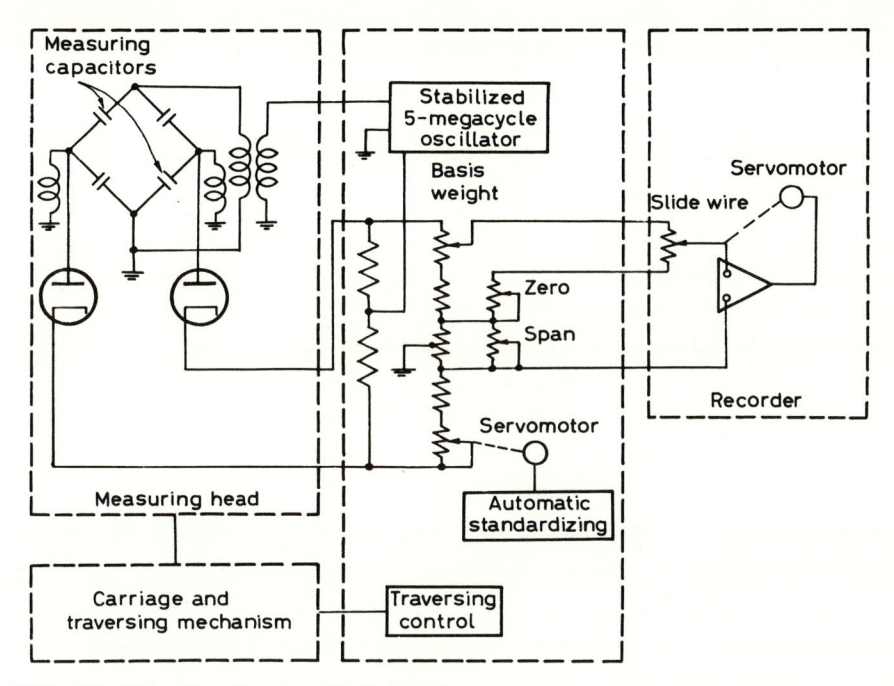

FIG. 10.23. The Honeywell Aquaset.

signal as the other. The Aquaset has several ranges of percent
moisture (0-10, 0-15, 0-20, and 30-50), and is said to detect variations as low as 0.01% on the low range. The admittance of a sheet
is a function of the sheet composition at virtually all frequencies.
Table 10.5 shows the 5-to-1 range of admittance for four typical
papers at a frequency of 200 Hz and with 8% moisture content. This
variation is undoubtedly due to a combination of factors such as the
chemical composition of the fibers, the morphology or structure of
the sheet, and the effect of added mineral matter. The operating
frequency of the measuring device should be a value that provides
good signal sensitivity for the grade of paper concerned. If this
frequency is not chosen correctly, variations in filler content, in
type of fiber, and in coating weight appear as noise in the signal.
In extreme cases where these quantities fluctuate widely, the process
variation may obliterate the desired signal.

TABLE 10.5

Relative Admittance of Different Types of Papers
at 200 kHz and 8% Moisture Content

1.	Coated book paper	1.00
2.	Kraft paper	0.35
3.	Bond paper	0.29
4.	Newsprint	0.19

A few remarks can be made regarding the comparative merits of electrical resistance and the dielectric-loss methods of measuring the moisture content of paper. The former appears to be more reliable when the moisture content is high and the resistance is low than it is for low moisture contents. The measurement of very high resistance of paper has been extended to the safe limit in the Hart moisture meter. The dielectric-loss method, on the other hand, appears to give its most reliable reading on low-loss, high-resistance paper of low-moisture-content sheets. If too much loss is introduced in the oscillating circuit, it becomes unstable. Adjustment of the circuit constants of the Moisture Register, however, should enable it to be extended to cover almost any range. For instance, the tightness of coupling of the contactor to the resonant circuit would control the effectiveness of the moisture content in decreasing the amplitude of oscillation.

5.2.4 Microwave Method. A general-purpose microwave moisture meter has been developed by Taylor [29]. The theory and design features of this type of moisture meter have been described in Chap. V and in a number of publications [30-37] and the reader is referred to them for a detailed study of the principles and the circuitry involved. The microwave moisture meter has been applied for the measurement of moisture content of paper in several paper mills. This moisture meter and the sensing element differ from others which have been previously discussed in that it operates at all moisture levels and can be applied at the press-section and the first part of the dryer section as well as for feedback from the dry end. The operating frequency of the moisture meter is 22.235 GHz, the response time is in the nano-second range, and the mode of transmission is through surface waves on single solid conductor detectors such as silicon diodes.

The system detects an absorption resonance corresponding to the transition between two rotational energy levels of the water molecules. It appears that there are few if any other substances with energy transitions corresponding to this particular resonant frequency, so the interferences encountered in, say, infrared absorption measurements are not present. The gain factor of sensitivity to water as against sensitivity to dry fiber exceeds 200 to 1 at high moisture contents, and is still about 50 to 1 at 3% moisture. Coating and filler minerals and variations in fiber composition do not seem to interfere to a greater extent than 0.25% moisture. This suggests an inherent signal-to-noise ratio of 32 to 1, far better than any of the other systems discussed so far. However, it has been reported that unexpected thermal noise disturbances due to temperature variations in the field are observed in field applications. Since the device detects only free water, the signal attenuation of microwave energy is linear with percent moisture only at relatively high values (over 10%). Below this level the relationship follows a square law curve; but this square law relationship appears to be consistent and identical for all types of paper, and it can be readily taken into account in the design of the readout device.

5.2.5 Electrolytic Method. An instrumental analytical technique for the determination of water in the range of 11 to 200 millimicrograms in small paper samples, e.g., 0.6 mg, has been developed to provide a rapid and easily applicable means for analyzing selected specific areas of sheets and paper products. The electrolytic measurement of water on which this procedure is based has been reported by Keidel [37], Taylor [38], and Cole et al. [39]. This principle has been adopted by Armstrong et al. [40] for the measurement of paper samples. The water in the sample is vaporized in a microoven by a controlled heating program. The released water vapor is carried by a stream of dry nitrogen into the electrolysis cell, where the water is absorbed by a thin, continuous film of anhydrous phosphoric acid located between two platinum helical electrodes. The absorbed water is electrolyzed by impressing a potential on the electrodes.

The signal from the electrolysis instrument is proportional to the current passed during the electrolysis and is recorded on a strip chart, while the area under the signal-time curve is simultaneously integrated by a digital readout integrator. Thus, both the area under the plotted curve and the total integrated count are a measure of the total amount of water driven out of the paper sample. A complete determination can be achieved in as little time as 10 min by this method, in contrast to the 4 to 5 hr required by standard oven-drying and weight-loss techniques. The electrolysis process is essentially specific for water; only certain basic materials, such as ammonia and some organic compounds, are possible interferences. Also, materials which can be absorbed by the phosphoric anhydride and undergo an oxidation or reduction under the test conditions will give erroneous results. Fortunately, such interference possibilities rarely are encountered in paper testing. The electrolytic determination of water has particular advantage over a weight-loss method for the treatment of samples which, on heating, may lose other volatile components in addition to water.

The apparatus of Armstrong et al. [40] shown in Fig. 10.24 consists of a tank of dry nitrogen, a magnesium perchlorate drying tube for the incoming nitrogen stream, a brass microoven, a Consolidated Electro-dynamics Corporation Model 26-301 Moisture Monitor, a flowmeter, a Minneapolis-Honeywell Model 51R10 integrator, and a Model G11A Varian recorder. The details of the microoven employed to vaporize the moisture in the paper samples are shown in Fig. 10.25. The microoven heater is regulated by an automatic timer and an adjustable voltage control. The output signal from the Moisture Monitor is recorded by the 10-mV Varian strip chart recorder and is simultaneously integrated by the Minneapolis-Honeywell integrator,

5 MOISTURE IN PAPER AND PAPER PRODUCTS

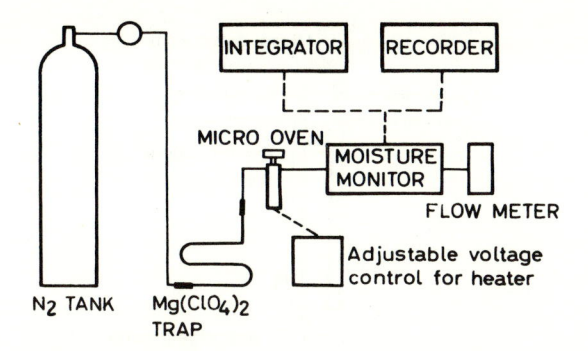

FIG. 10.24. Armstrong electrolytic moisture meter.

which has a range of 0 to 1000 counts/min corresponding to a signal output range of 0 to 10 mV.

The calibration curve for quantities of water ranging from 11 to 184 mg was made from a series of 49 points of integrator count vs micrograms of water contained in water-equilibrated nitrogen samples ranging in volume from 0.5 to 8.5 ml. The equation for the best straightline fit of these data was calculated by the method of least squares to yield:

$$y = 155.70 + 11.195x \qquad (10.4)$$

where x is micrograms and y is the integrator count.

To report percent moisture in paper by the electrolysis system, it is necessary to know the initial sample weight. As a microbalance weighing of each sample for every determination would defeat the simplicity desired, a series of uniformly punched paper samples were weighed to determine whether or not the percent weight variation of such samples would be within limits compatible with the established precision of the electrolytic method. A series of eight 1/8-in. diameter disks punched out of a commercial 16-point white tag board, conditioned at 43% relative humidity, were found to have an average weight of 2.652 mg. The standard deviation of these weighings was 0.083 mg or 3.1%. As this variation is less than the variation in

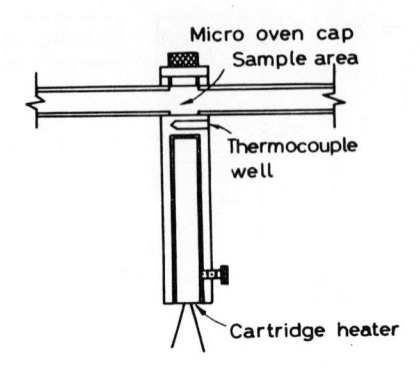

FIG. 10.25. Schematic diagram of microoven.

the actual moisture determination, it is feasible to make a series of moisture determinations on the basis of a predetermined average weight for any given paper sample. The moisture contents of the 100 punched disk samples were determined simultaneously by the electrolytic method and by their weight loss after treatment in the microoven. Table 10.6 shows the results obtained for the entire range of samples studied by Armstrong et al. [40].

5.2.5 Infrared Method. The theory of the infrared absorption of water and its utilization for the measurement of traces of moisture content in different materials has been described in the literature [41-45] and Chap. VI. A slightly different device based on the differential absorption of infrared radiation on two neighboring wavelengths, one in a water resonance band and the other just outside it, has been developed (GE Company). This device uses the water absorption peak (in cellulose) of 1.94 μm as the measuring channel, and the adjacent band of 1.82 μm as the reference channel. The sheet is exposed to narrow band radiation at these two wavelengths, in rapid alternation through the action of a chopper disk equipped with spikepass filters (Fig. 10.26). Backscattered radiation collected in an integrating sphere activates a lead-sulfide detector. The signals are amplified, demodulated, and subtracted, and the result is recovered.

5 MOISTURE IN PAPER AND PAPER PRODUCTS   709

TABLE 10.6

Comparison of Moisture Content for 15-Point White Tag Board as
Determined by Electrolysis and Weight Loss to Establish
Operating Range and Precision of the Method

	Sample diameter (in.)[a]	Average wet weight for 10 samples (mg)	Av. water weight (mg) By electrolysis	Av. water weight (mg) By weight loss	Av. percent water, ± By electrolysis	Av. percent water, ± By weight loss
A	3/32	1.479	0.004±0.001	0.030±0.009	0.24±0.06	2.03±0.63
	1/8	2.506	0.015±0.001	0.031±0.010	0.60±0.03	1.24±0.40
B	3/32	1.528	0.081±0.004	0.094±0.011	5.27±0.09	6.13±0.60
	1/8	2.662	0.159±0.007	0.152±0.018	5.96±0.08	5.72±0.71
C	3/32	1.560	0.130±0.010	0.132±0.013	8.31±0.20	8.44±0.68
	1/8	2.803	0.254±0.014	0.245±0.018	9.06±0.27	8.74±0.38

[a] A are samples conditioned over $CaSO_4$, approximately 0% relative humidity.

B are samples conditioned over saturated solution of K_2CO_3, approximately 43% relative humidity.

C are samples conditioned over saturated solution of NH_4Cl, approximately 79% relative humidity.

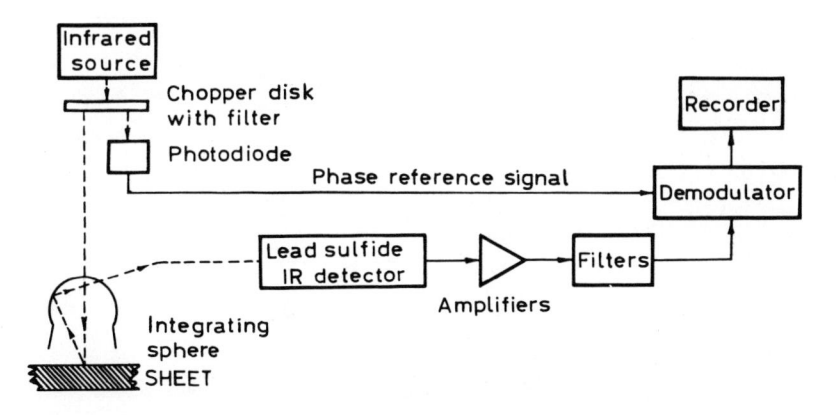

FIG. 10.26. Infrared (backscatter) moisture meter).

The amount of backscatter in both bands depends on the basis weight of the sheet, the degree of bonding between fibers, and the chemical constitution of both fibers and nonfibrous sheet constituents. However, because of the sharp moisture peak at 1.94 μm, the difference between the reflectance values at these two wavelengths is almost solely a function of the moisture content. If calibrated for basis weight or provided with a compensating signal, the backscatter meter should give a moisture reading that is practically unaffected by electrolytes, fillers, or coating variations. Its most precise application is limited to no more than 12% moisture content and to grades of paper not exceeding 70 lb/3300 ft^2 in basis weight.

5.2.6 Distillation and Titration Methods. Besides the standard oven-drying method and the electrical methods described earlier, there are a few other well-known methods which have been used for the measurement of moisture content of paper and paper products. These are distillation and titration methods. The distillation method as originally developed by Dean and Stark [9] and considerably improved by Bidwell and Sterling [46] has been fully described in Chap. III.

Titration methods, such as those based on the reaction of water with acetic anhydride and with the Karl Fischer reagent, appear most promising in the presence of volatile materials. The latter is especially to be recommended. The method gives the same results as the oven-drying procedure for paper containing no volatile substances and gives more reliable results when volatile matter is present. These two titration methods have been elaborately described in Chaps. III and IV.

6 CONCLUSION

The measurement of the moisture content of wood, paper, and bagasse is beset with a number of difficulties. Most of the important problems related to moisture measurement result from limitations

6 CONCLUSION

in the electrical or electronic methods of moisture measurement. However, the electric or electronic moisture meters provide the only practical means for moisture inspection and control, because other reliable methods are destructive and time consuming. The limitation of electric moisture meters is their inability to measure reliably moisture levels above the fiber saturation point. According to Thode [28], at the fiber saturation point (FSP approximately 30% moisture content) perhaps one quarter of the removable water is weakly bound to the cellulose by dispersion forces or hydrogen bonds, and has therefore lost its translational and rotational degrees of freedom. At the same time, the hydroxyl groups on the ends of cellulosic chains possess much more vibrational energy than they do when the sheet is dry.

The equilibrium between bound and free water depends on both temperature and the concentration of free water. Thus, as the sheet is dried below the FSP and the concentration of free water drops, the relative amount of bound water also decreases. The water molecules so released gain rotational and translational degrees of freedom, while cellulose chain groups to which they were bound now form bonds with one another and lose their vibrational degree of freedom. The result of this interplay is a nonlinearity in the relationship between electronic properties and moisture content, so that any single electronic measurement of rf impedance, microwave or infrared absorption is not directly proportional to moisture content in such a situation. It is therefore necessary to take the average of three readings at least.

New techniques utilizing nuclear properties of the hydrogen atom contained in water offer excellent possibilities of moisture measurement in bagasse, wood, and paper as well. These methods of moisture measurement are known as (1) the neutron scattering method, (2) backscatter of gamma radiation, and (3) the nuclear magnetic resonance method. The neutron scattering method has been originally developed [47-49] for the measurement of moisture content of soil, concrete, or other solid inorganic materials and has been described in detail in Chap. VII. The second method depends on the measurement of attenuation or backscatter of gamma radiation and may be used to determine the gross density of a material. If the density of the dry material is known or can be determined independently, moisture content can be calculated from such radiation measurements. A limited amount of research has been directed toward evaluation of this approach to the measurement of moisture content of wood [50] as well.

It is possible to have an accurate measure of the moisture content of the material by using the nuclear magnetic resonance method. By this technique, the hydrogen in adsorbed water can be distinguished from hydrogen in carbohydrates by differences in the frequency dispersion of the resonances of the two types of hydrogen. However, the high cost of the apparatus and the involved techniques preclude this method for any but fundamental laboratory use. This method was originally developed for the measurement of moisture content of grains, cereals, and similar agricultural products [51-56], and therefore has been described in detail in Chap. XI.

Finally, the dielectric properties of these materials at certain microwave frequencies are theoretically promising indicators of moisture content, principally because the effect of water may overshadow the effect of other variables such as density and fiber packing and composition. This method was originally developed for the measurement of moisture content of building structures and building materials and has been therefore described in detail in Chap. XII. However, with certain modifications in the sampling procedures, it can be applied to bagasse, wood, and paper as well, as shown in the earlier descriptions.

REFERENCES

1. E. C. Lathrop, Ind. Eng. Chem., 22, 449 (1932).
2. D. F. J. Lynch and M. J. Goss, Ind. Eng. Chem., 23, 1249 (1932).
3. J. B. Molitor, Sugar J., 11, 3 (1949).

REFERENCES

4. H. Kumagawa and K. Shimomura, *Z. Angew. Chem.*, 36, 414 (1923).

5. Anonymous, *Bull. Impl. Inst.*, 27, 1 (1929).

6. A. Valenzuela and A. P. West, *Philippine J. Sci.*, 40, 275 (1929).

7. E. C. Munro and W. S. Wise, Proc. Meeting British West Indies Sugar Technologists held in Jamaica, 1960, p. 176.

8. Schmidt, Wiggins, and Yearwood, Proc. B.W.I. Sug. Tech., 1951, p. 163.

9. E. W. Dean and D. D. Stark, *Ind. Eng. Chem.*, 12, 486 (1920).

10. A. Pande and S. P. Agarwal, *Instrum. Pract.*, 20, 915 (1965).

11. R. H. Cotton, W. A. Harris, L. P. Orleans, and G. Rorabaugh, *Anal. Chem.*, 24, 1498 (1952).

12. American Society for Testing and Materials — Methods of Test for Moisture Content of Wood, ASTM Designation D2016-62T (1962).

13. Indian Standards Specification IS 287-1960. Maximum Permissible Moisture Content of Timber Used for Different Purposes in Different Climatic Zones.

14. A. Pande, *Lab. Pract.*, 12, 432, 661, 741 (1963).

15. R. L. Young and W. L. James, Control and Measurement of Moisture in Wood, *Humidity and Moisture*, 4, Reinhold, New York, 1965, p. 307.

16. F. Kollman and G. Hockele, Kritischer Vergleich einiger Bestimmungsverfahren der Holzfeuchtigkeit, *Holz Roh Werkstoff*, 20, 461 (1962).

17. A. J. Stamm, *Ind. Eng. Chem., (Anal. Ed.)*, 2, 240 (1930).

18. M. A. Rehman, J. Kishen, and B. I. Bali, *Indian Forest Bull.*, 213, 4 (1956).

19. W. L. James, U.S. Forest Products Laboratory Report No. 1660.

20. J. A. Shaffer, *Process Control Automation*, 1962, p. 501.

21. W. Brecht, Specifications Regarding the Moisture Content and Moisture Absorption of Paper, *Textilebrichte*, 29, 1 (1948).

22. K. W. Hardacker and R. D. Rawcliffe, *Tappi*, 35, 1684 (1952).

23. H. W. Verseput, Tappi, 34, 572 (1951).

24. J. A. Van den Akker and K. W. Hardacker, Tappi, 35, 138A (1952).

25. C. Delevanti, Jr. and P. B. Hansen, Paper Trade J., 121, 35 (1945).

26. C. R. Calkins, Tappi, 33, 278 (1950).

27. J. E. Fielden, Proc. Tech. Section, Paper Makers Assocn., U.K., 28, 513 (1947).

28. E. F. Thode, Control Eng., 67 (1964).

29. H. B. Taylor, A.E.I. Eng., 3, 66 (1965).

30. A. Watson, British Pat. 697,956 (1962); Building Research Station Report A78 (1961) and A93 (1957).

31. A. B. Bronwell and R. E. Beam, Theory and Application of Microwaves, McGraw-Hill, New York, 1947.

32. W. Gordy, Rev. Mod. Phys., 20, 668, 717 (1948).

33. J. B. Hasted and M. A. Shah, Brit. J. Appl. Phys., 15, 825 (1964).

34. P. O. Vogelhut, Nature, 203, 1169 (1964).

35. A. D. Ince and A. Turner, Analyst, 90, 692 (1965).

36. A. Pande, J.S.I.R. (India), 9A, 243 (1950).

37. F. A. Keidel, Anal. Chem., 31, 2043 (1959); U.S. Pat. 2,830,945 (1958).

38. E. S. Taylor, Refrig. Eng., 64, 41 (1959).

39. L. G. Cole, M. Czuha, and R. W. Mosley, Anal. Chem., 31, 2048 (1959).

40. R. G. Armstrong, K. W. Gardiner, and F. W. Adams, Anal. Chem., 32, 752 (1960).

41. A. Finch, P. N. Gates, K. Radcliffe, F. N. Dickson, and F. F. Bentley, Chemical Applications of Far Infrared Spectroscopy, Academic Press, New York, 1970.

42. E. F. H. Brittain, W. O. George, and C. H. J. Wells, Introduction to Molecular Spectroscopy, Academic Press, New York, 1970.

43. S. S. Pinchas and I. Laulicht, Infrared Spectra of Labelled Compounds, Academic Press, New York, 1971.

REFERENCES

44. W. E. Keder and D. R. Kalkwarf, Anal. Chem., 38, 1288 (1966).

45. E. G. Mahadevan, Analyst, 92, 717 (1967).

46. G. L. Bidwell and W. F. Sterling, Ind. Eng. Chem., 17, 147 (1925).

47. J. W. Holmes and A. F. Jenkinson, J. Agr. Eng. Res., 4, 100 (1959).

48. J. W. Holmes and K. G. Turner, J. Agr. Eng. Res., 3, 199 (1958).

49. P. G. Marias and W. B. DeV. Smith, S. African J. Agr. Sci., 3, 115 (1960).

50. P. Kajane and A. Hollming, Paperi Puu, 40, 153 (1958).

51. W. L. Rollwitz, Nuclear Magnetic Resonance as a Technique for Measuring Moisture in Liquids and Solids, Humidity and Moisture, 4, Reinhold, New York, 1965, p. 149.

52. W. L. Rollwitz, Proc. Nat. Electron. Conf., 12, 113 (1956).

53. C. H. M. Van Bavel, Proc. Intl. Sym. Humidity and Moisture, Washington, D.C., 171 (1963).

54. D. A. Lane, B. B. Torchinsky, and J. W. T. Spinks, Determining Soil Moisture and Density by Nuclear Radiations, Proc. Symp. Use of Radioisotopes in Soil Mechanics, by American Society for Testing Materials (1952).

55. J. W. Holmes, Aus. J. Appl. Sci., 7, 45 (1956).

56. A. I. Johnson, Methods of Measuring Soil Moisture in the Field, by U.S. Government Printing Office, Washington, D.C. — Geological Survey Water Supply Paper 1619-U, 1962.

Chapter XI

MOISTURE IN FOODS AND ALLIED AGRICULTURAL PRODUCTS

1 INTRODUCTION

Moisture plays a very important role in the storage and processings of different kinds of foods and agricultural products. Under practical storage conditions, moisture content is usually the principal governing factor as regards the keeping quality of cereals, grains, and other agricultural products. Near the critical moisture content small differences in moisture content cause large differences in keeping quality. The phenomenon of moisture sorption and its measurement in foods is a complicated one, because these materials are a complex mixture of organic substances. Acidity values and other indexes of deterioration have been determined by Zeleny and Coleman [1] for samples of wheat having sufficiently high moisture content, i.e., above the critical moisture content. The changes on percentage basis that occurred in acidity are illustrated graphically in Fig. 11.1. It is obvious from the data that the fat acidity, the phosphate acidity, and the total titrable acidity increase as the wheat deteriorates in storage.

718   XI. MOISTURE IN FOODS

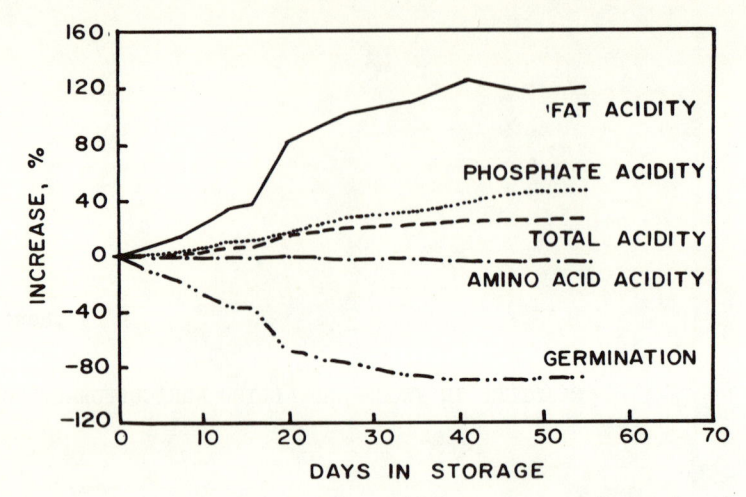

FIG. 11.1. Percentage change in acidity values and germination of hard red winter wheat stored at 15.35% moisture content.

The biological and chemical effects produced in grains and seeds during the process of storage have been studied by Hukill [2]. According to him the biological and chemical responses of grains to moisture are closely related to their responses to temperature. Moisture contents of from 10 to 14% depending on temperature, length of storage period, and other factors are required for keeping grain in good condition during storage. Grain that contains too much water is subject to rapid deterioration from mold growth, heating, insect damage, and sprouting.

The scope of this chapter is very wide and diverse, as it covers the measurement of moisture content of cereals, grains, flours, stimulants, beverages, fruits, sugar and sugar products, dairy products, etc.

2 SORPTION OF MOISTURE BY CEREAL GRAINS

All cereal grains are hygroscopic, i.e., when they are exposed to an atmosphere equilibrium is attained. For any given level of

moisture content in the grains, there is a corresponding level of
atmospheric humidity at which the net moisture exchange would be
zero and moisture equilibrium would exist. Moisture is believed to
exist in several forms in agricultural products. The two forms we
are concerned with are free water and bound water. Free water is
best defined as water that is capable of acting as a solvent and is
found in the capillaries and other interstitial spaces of the grain
structure. Bound water is electrostatically bound into the complex
structure of the large molecules of carbohydrates, proteins, starch,
and other colloidal substances, and is physically although not chemically a part of the structure. However, because of hydrogen bonding
there is no distinct line of demarcation between free and bound water.

A very useful method to study the moisture sorption characteristics of hygroscopic materials, such as cereal grains, is by means
of isotherms which are obtained by plotting moisture sorption values
against the different relative humidities at a constant temperature.
Predominant constituent substances occurring in cereal grains are
starch and proteins which govern the interaction of these substances
to water. One theory currently used by some investigators to explain
observed rates of moisture loss is that the rate of flow is a function of the dynamic equilibrium moisture content, a changing base to
which the moisture content of the grains is continually approaching
at a predictable rate. Other studies are interpreted to show that
the rate of flow is proportional to the square of a potential whose
base is the static equilibrium moisture content.

It is useful to consider the physical relations of grain with
water vapor, such as hygroscopicity, rates of moisture exchange and
definition and measurement of moisture content. Babbitt [3], by
changing the temperature gradient, showed that when the moisture
concentration gradient is in one direction and the vapor pressure
gradient in the opposite direction, the moisture flow is from the
region of high vapor pressure to that of low vapor pressure. Typical
equilibrium moisture curves or sorption isotherms at three different
temperatures are shown in Fig. 11.2. These curves are for grain

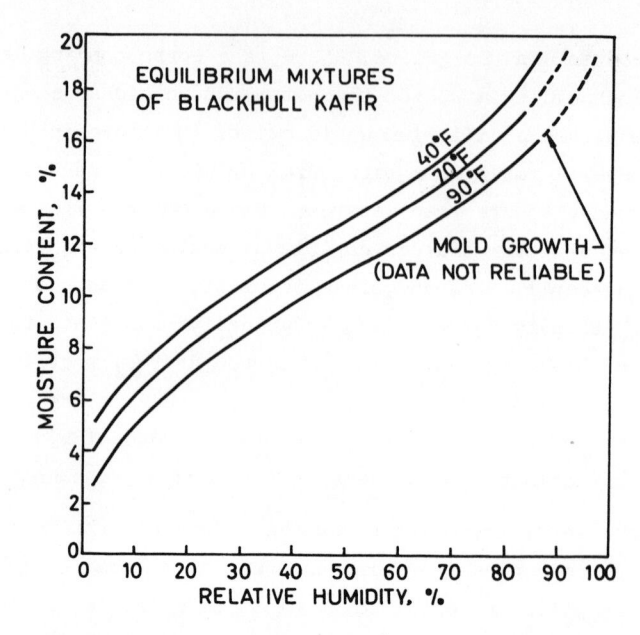

FIG. 11.2. Sorption curves for grain (sorghum) at three different temperatures.

sorghum as presented by Fenton [4]. Curves for other grains would fall close to those for sorghum, the grains with high oil contents tending to fall further to the right. It has been observed that the equilibrium moisture curve is influenced by the direction from which the grain approaches equilibrium. Grain approaching equilibrium from a lower moisture content will not reach as high a moisture level as when it approaches from a higher moisture. This effect, called hysteresis, is illustrated in Fig. 11.3, which shows adsorption curves for corn according to Hubbard et al. [5]. From the isotherm by using the well-known Clayssius-Clayperon equation, it is possible to calculate the heats of sorption (integral and differential) as well as changes in free energy and entropy as a result of moisture sorption. Shedd and Thompson [6] made estimates of the heat of vaporization of the water in grain from this relationship by using experimental observations on equilibrium moisture contents.

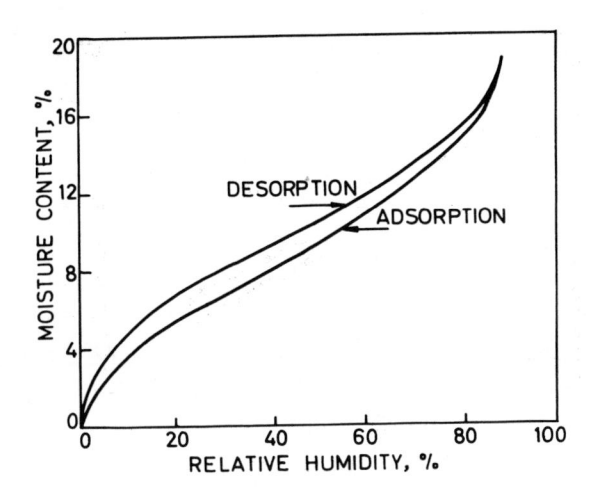

FIG. 11.3. Adsorption and desorption curves of corn illustrating hysteresis.

It is now well known that the optimum moisture content of many such foods are close to the water vapor monolayer capacity measured by the BET adsorption isotherm. It may be that the monolayer in these cases represents a condition in which one water molecule is attached to each accessible polar group in the protein, fat, and carbohydrate of the foods. The nonfreezing water in several food proteins is in fact equivalent to about two molecules per peptide group. This water may be necessary to prevent interaction between adjacent carbohydrate or protein molecules, which would interfere with later reconstitution of the food by water; and it may also offer just sufficient protection against attack on the polar group of oxygen.

According to the BET theory [7] of multilayer adsorption, the first portion of the isotherm represents the adsorption of the first layer of water vapor onto the surface of the adsorbing material; the region of inflection represents the deposition of a second layer of water molecules; and the final curved portion represents the continued adsorption of additional layers. In each of the three segments of the isotherm a different relationship between vapor pressure and moisture content exists. In the initial portion of the isotherm, the vapor pressure-moisture content relationship is governed by the binding energy between the water molecules and the adsorbing surface. The binding energy depends on the physical structure of the surface and its chemical constitution and on the physical and chemical properties of water. The magnitude of the binding energy is thus the resultant of various effects. The extent to which the isotherm is displaced toward the moisture content axis is an indication of the binding energy between water and the adsorbing substances.

2.1 Measurement of Moisture Content of Cereal Grains

2.1.1 Introduction. A large number of methods and techniques which have been developed and applied for measurement of moisture content of cereals, grains, and similar agricultural products are the gravimetric, distillation, Karl Fischer, gas chromatographic, electrical or electronic, infrared spectroscopic, and NMR methods.

These methods can be broadly classified into direct (chemical) and indirect (physical) methods. The direct methods are capable of giving very accurate results but are complicated and time consuming, whereas the indirect methods are instantaneous and by their use continuous measurement or monitoring of moisture content can be made. They, however, require calibration against the chemical methods, but once calibrated can be relied upon to give accurate results.

2.1.2 Gravimetric Methods. Oven-drying and similar methods in which weighing is involved can be broadly classified as gravimetric methods. The oven-drying method has undergone considerable improvements and at present semiautomatic oven-drying equipments are available commercially which have built-in balances for weighing dried samples in hot condition without removing them from the oven. Examples of such rapid automatic moisture testers are the Cartersimon moisture tester mostly used in the U.K. and Europe and the Brabender semiautomatic moisture tester mostly used in the US and Canada. The balances are calibrated directly in terms of moisture content. This method avoids errors in weighing due to weighing in an external balance, which has been found [8] to be as much as 0.4% in the case of 500-g sample. Factors involved in the oven-drying method and the sources of errors have been discussed by a number of researchers [9-12], and in Chap. II.

The methods for determining moisture content as specified in the Official Grain Standards of the United States and in the United States Standards [13] of beans, peas, lentils, and rice are all air oven methods. It is recommended in the specifications that the grains should be ground in a ball mill and then dried at 130°C for 1 hr in an air oven. The grinding of grain samples for moisture determination has been the subject of much controversy. Some investigators are of the opinion that moisture changes during grinding, i.e., loss of moisture to the air or sometimes gain of moisture from the air may take place. This danger is increased when the moisture content is high and when large numbers of samples are ground consecutively in a burr mill which rapidly becomes hot. However, it has been shown by Zeleny and Hunt [14] that there is no loss or gain of moisture when samples having 16% or less moisture are ground on the Wiley Intermediate Mill with an 18-mesh sieve.

Another commonly used air oven method for determining moisture in grain provides for heating a weighed portion of the finely ground grain for 2 hr at 135°C. This method gives slightly higher results than the previously described method and is an official method of analysis for cereal grains of the Association of Official Analytical Chemists [15]. Recently specifications for the determination of moisture content of cereals and cereal products have been made by International Standards Organization and is designated as Draft Specifications ISO 909. In these specifications full details of the grinding procedures and drying methods are described.

Various modified-air oven methods using special heating equipment have been devised to shorten the time required as compared

with standard oven methods. In general these methods provide for
heating the material to considerably higher temperatures than those
employed in the usual oven methods. Heating may be accomplished by
ordinary electric radiators [16], or by means of a high-frequency
and a high-voltage field. Out of all these sources of heat radiation, heating by infrared has been found to be most satisfactory, as
the materials can be bone-dried in less than 15 min by this radiation
without raising the temperature of the sample beyond 80°C.

One of the most widely accepted methods for determining the
moisture content of grains is based on drying the finely ground
grains at a temperature of 98 to 100°C in an oven chamber maintained
at a pressure of 25 mm or less of mercury. Heating is continued for
about 5 hr until no appreciable further loss of weight occurs. This
method is one of the official methods of the Association of Official
Analytical Chemists [15] for grains and gives results in reasonably
good agreement with the 2-hr, 130°C air oven drying method. At the
present time the vacuum oven method used by most of the industrial
laboratories is similar to that recommended by the Association of
Official Analytical Chemists. According to this method the material
whose moisture content is to be determined is heated in the vacuum
oven at 70°C for 6 hr at a pressure not exceeding 100 mm Hg. Makower
et al. [17] have utilized the vacuum oven method very successfully
for the measurement of the moisture content of dehydrated vegetables
by drying the samples in the vacuum oven at 60°C for 60 hr to get
constant weight. Another method to keep the sample thermally stable
is the drying in a gaseous atmosphere. Snyder and Sullivan [18]
have employed heated nitrogen and hot air drying technique for determining moisture content of wheat flour.

2.1.3 Distillation Method. Two distillation procedures that are
classed as standard methods warrant discussion. The first of these
is the Brown-Duvel method. For many years this was the standard
moisture method for grain inspection in the United States [19-21]
and Canada [22]. It is also listed as a method for cereals by the

American Association of Cereal Chemists. The second standard distillation method is the toluene or benzene distillation. It is an official method of the Association of Official Analytical Chemists for grains and stock feed.

The Brown-Duvel distillation method [21] has found wide applications in testing of cereals and grains. The procedure to be followed for each kind of grain is described in detail in US Department of Agriculture Procedure [22]. The equipment generally makes provision of heating six flasks simultaneously; four-, two-, and a single-unit equipment is also available. A 100-g sample of whole grain is heated in the flask with 150 ml of nonvolatile oil to a specified cutoff temperature (180°C for wheat). The hinged heating unit is then lowered to a predetermined position so that the oil-grain mixture cools to 160°C. The amount of water that is distilled into the graduated cylinder is read in millimeters and reported as percent moisture. The determination takes about an hour and with a suitable number of equipments, one can make about 12 to 18 determinations/hr. The method has certain advantages. It can be readily followed by inexperienced operators. Only a relatively crude torsion balance is required to weigh 100 ± 0.1 g. The remainder of the equipment is reasonably rugged or easily replaced. The Brown-Duvel method can thus be used in elevators where other methods involving the weighing of small samples on fine balances would not be feasible. However, the apparatus used for this purpose must be standardized to provide a definite amount of heat in a definite period of time. The method is highly arbitrary and the exact procedure to be followed must be established for each kind of grain in order to obtain results equivalent to those obtained by the applicable official oven methods. The criticism that the Brown-Duvel [21] method underestimates moisture content may be an exaggeration. The method was originally standardized against the water oven, a procedure which is still used for corn, although for other grains it is standardized against the vacuum oven or other methods as well, as shown in Fig. 11.4.

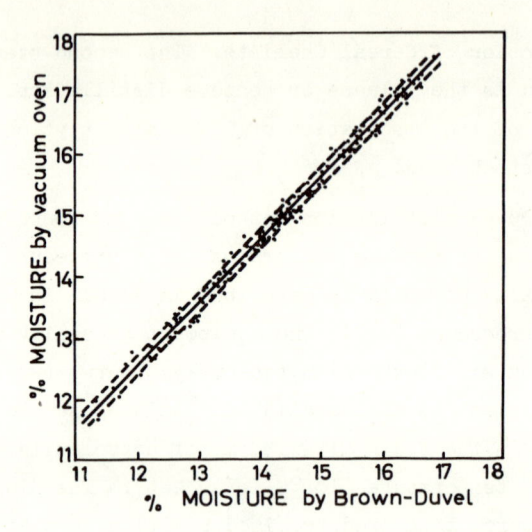

FIG. 11.4. Relation between vacuum oven and Brown-Duvel moisture determinations.

The other distillation method is known as an azeotropic method since water and toluene or benzene form an azeotropic mixture. One of the official methods of the Association of Official Analytical Chemists [15] provides for boiling the finely ground grain in toluene in an apparatus that condenses the volatilized materials, collects the condensed water in a graduated tube, and returns the condensed toluene to the boiling flask. The boiling is continued as long as any water accumulates in the graduated tube and the moisture content of the grain is calculated from the volume of water condensed. Since the boiling point of toluene is 111°C, all substances that boil at or below this temperature, including all water that will be released at this temperature, are distilled. Also, since the water is measured volumetrically, no water-insoluble volatile material can be measured as moisture content. Detailed specifications using distillation techniques for moisture determination of cereals and cereal products have been drawn up by the Corn Industries Research Foundation [23]. According to these specifications a sample of 20 to 30 g of corn, ground in a small Wiley mill with 20 sieve, is introduced into the flask with 75 ml of toluene. Distillation is started slowly (1 to 2 drops/sec), but is speeded up (2 to 4 drops) after the first half-hour, and is continued for 48 hr. Moisture collects in the graduated trap from which the excess toluene flows back into the flask. Careful attention to specified details of technique and standard apparatus are required for accurate results. Toluene distillation is used for corn, corn gluten meal and benzene distillation is used for corn gluten feed and sweetened feed. Though the method is comparatively slow, it should have wide applications for cereals and cereal products and has much to recommend it as a basic reference procedure.

2.1.4 The Karl Fischer Method. Because of certain practical difficulties in its application, the Karl Fischer method has been sparingly used for determining moisture in grain, although Fosnot and Hamon [24] have applied the method to wheat and barley, who recommend that cereal products and similar materials should be ground to about 0.5-mm particle size and warmed with methanol, excluding atmospheric moisture as far as possible. After cooling, a slight excess of reagent should be added, the excess being back titrated with a standard solution of water in methanol. Hart and Neustadt [25] have successfully adapted the Karl Fischer method to all cereal grains and have used it to test the accuracy of official oven methods. Because of the technical complications and the time-consuming nature of this analytical method, its usefulness for routine measurement is likely to remain quite limited. However, it is one of the most accurate methods available and serves as a standard for comparing the accuracy of other methods. It is also used for calibrating electric and electronic instruments.

2.1.5 Gas Chromatographic Method. In the technique devised by Weise et al. [26] gas chromatography is used to separate and analyze a methanol-water mixture extracted from the moisture-containing material. Accurate to better than one part in a hundred of the water content of representative materials, the method is not designed for field use but should provide good laboratory control and a means for standardizing equipment used for field testing. The technique of gas chromatography is well suited for the determination of water in a variety of food products, especially high-moisture food products, which are difficult to analyze by other procedures. A suitable gas-liquid chromatographic method can satisfy most criteria of a good quality control procedure, namely speed, accuracy, and simplicity.

In the development of the gas chromatographic method for moisture determinations it was necessary to find a suitable support medium and partitioning agent that would produce methanol and water peaks with adequate separation and suitable shape. Such a combination is provided by commercial 30 to 40 mesh polytetrafluoroethylene coated with a polyethylene glycol partitioning agent. The peak areas for known methanol-water solutions are measured by means of a planimeter to standardize the chromatographic procedure. This standardization, once determined, should hold indefinitely. To determine the moisture in grain, a known amount of anhydrous methanol is added to a weighed amount of grain to extract the water from it. Ten µl of the extract are injected into the chromatograph for analysis. The relative amounts of methanol and water are recorded as peaks on a graph and provide a basis for the moisture determination. Recording the chromatogram and measuring the peak areas requires about half an hour. Extracting the water from the material requires additional time depending on the procedure used. If a wet grinding procedure is used, only a few minutes are required but if room-temperature extraction is employed, a day or more may be necessary.

The chromatographic method for moisture determination overcomes many of the shortcomings of earlier laboratory methods. Not only is it highly specific for water, but it is largely instrumental, thus eliminating difficult chemical techniques. Furthermore, it is free from the requirement of frequent restandardization, in contrast to the Karl Fischer titration method, for example. The method, however, does not eliminate the basic problem common to all known methods for water determination, namely, the definition of that which constitutes the actual water content of a given type of material. The new method furnishes a reliable value for water-extractable methanol, and there is a good reason to believe that this water consists of all but that which is chemically combined with the sample of grain. The techniques of the gas chromatography and gas-liquid chromatography have been described in some excellent books [27-30] and Chap. III and the reader is referred to these for further study of the subject.

Schwecke and Nelson [31] have developed a method for the determination of moisture content of foods by employing gas chromatography. The procedure described by these researchers utilizes a column prepared by coating Fluoropak 80 with 10% Carbowax 400. This column gives water a low retention time which allows rapid analysis. Secondary butanol is used as an internal standard and methanol is used to extract water from the product to be analyzed.

The reproducibility of the procedure was checked by preparing two mixtures of water, secondary butanol, and methanol equivalent to a mixture which would be obtained from 15-g samples of cereals containing 25 and 30% moisture, respectively. The reproducibility of the entire method is illustrated in Table 11.1. This method has been used for routine quality control work for over a year. The column is very stable if not heated above 130°C. These investigators have used one column continuously for a year with little change in resolution.

This method is specific for water in the sense that the presence of some other component in grain having exactly the same elution characteristics as water or methanol is highly improbable. When once the instrument has been calibrated, restandardization is not necessary and no special technical skill is required to operate the instrument. Multiple analyses of a sample as a function of various soaking times are possible on the same extract since only 10 µl of the original 50 ml are used for each sample injection; also, a permanent record of all analyses is obtained which, if desired, may be evaluated and/or reevaluated at some later date. The gas-chromatographic method can be considered as a basic laboratory method for determination of moisture content of cereals and grains and similar agricultural products, provided the extraction time is reduced to shorter periods. Hart and

TABLE 11.1

Reproducibility of Entire Procedure

Food products	Moisture (%) GLC procedure	Other methods
Cereal pellets	21.5, 21.7, 21.3, and 21.5	18.7 infrared balance
Dried raisins (no. 1)	9.2, 9.2, and 9.3	7.1 vac. oven (6 hr)
Dried raisins (no. 2)	13.6, 13.8, and 13.8	11.3 vac. oven (6 hr)
Flour (no. 1)	12.6, 12.6, 12.6, and 12.6	12.5 air oven (130°C)
Flour (no 2)	13.5, 13.7, and 13.6	13.5 air oven (130°C)

Neustadt [25] have reported that a 3-min dry grind in a modified Stein mill, followed immediately by a 5-min wet grind in methanol near its boiling point, effectively removes all water present in the grain. Accordingly, such a procedure might be a desirable means for providing the extraction required by the chromatographic method. In the chromatographic method an analysis time of approximately 18 min is required to obtain each chromatogram. With practice, 10 to 12 additional min are then needed to carefully measure the peak areas. Consequently, total analysis time is about half an hour, proportionately longer if more than one sample injection is desired. This technique offers an independent method for the determination of water in methanol extracts from grain and is useful in resolving the discrepancies which might arise between oven-dry and Karl Fischer determinations of moisture in grain. This method can be easily adopted for the moisture determination in many other types of materials in which the water can be quantitatively extracted by methanol.

2.1.6 Electrical or Electronic Methods. Based on the principle of dc resistance variation and capacitance variation with moisture content, a large number of dc conductivity- or resistance-type as well as dielectric or capacitance-type moisture meters have been developed in UK, USA, Europe, and India during the last 30 years. Geary [32] has given data about 12 types of electrical moisture meters manufactured in England. All meters introduce grain into an electrical circuit and measure the resistance and/or dielectric constant of the grain. The three resistance meters that appear most suited for grain are the Tag Heppenstall, the Universal, and the Marconi (Type TF 933). In the Tag Heppenstall meter the electrical resistance of the grain is measured as it passes between two corrugated steel rolls that serve as electrodes. One is motor driven (or hand driven) and the other idles. Grain is poured into the hopper and is partially crushed to provide good electrical contact as it passes between the rolls. A selector switch on the separate instrument box is turned to one of eight positions and the position of the galvanometer needle is read. The temperature of the grain must be known, and charts are provided to convert temperature and galvanometer readings to moisture content.

The Tag Heppenstall meter appears to be one of the most accurate instruments now available for measurement of moisture content of grains; the standard error of estimation for testing hard spring wheat is about ±0.23% [33-34]. It is certainly the most rapid; tests can be made in 10 to 20 sec. When it is required to determine whether the moisture content is over or under a specified value, tests can be made at a rate of 8 to 10/min. The meter has the further advantage that calibration charts are available for various types of wheat and for almost all other grains. The meter cannot, however, be used for flour and other cereal products.

In the Universal meter, a hand-driven megger establishes a voltage across a sample of pressed grain and electrical resistance is indicated on an ohmmeter of the dynamometer type. The instrument thus requires neither batteries nor power supply. A test is made on a 20-g sample weighed to the nearest kernel. This is transferred to a steel cup with a cylindrical plastic lining and an internal diameter of 1 5/8 in. The cup is placed under the ram of a screw press operated with a ratchet lever, and the grain is pressed to specified thickness (0.40 in. for wheat), which is indicated by a micrometer device on the instrument. The megger is then cranked rapidly and the meter is read. An armored thermometer is provided for reading the temperature of the grain. The test requires about 1.5 min, which is also about the time required for tests on most electrical meters other than the Tag Heppenstall. A circular slide scale for converting temperature and meter readings to moisture content is provided on the front of the instrument. By adjusting the weight of the sample and the thickness to which it is pressed, the same conversion scale is said to be accurate for all grains. Investigation of a prototype of the Universal meter showed [35] that it is almost as accurate as the Tag Heppenstall meter. The standard error of estimate was found to be ±0.28% [35]. If the instrument is accurately calibrated for each cereal and for different temperatures, it has several advantages. The calibrated meter can be used for a wide variety of seeds, including flax and mustard, without modification.

The Marconi (Model TF 933) is a battery-operated instrument of the resistance type. It has been described in detail by Brockelsby [36]. The test cell is based on a C-clamp holding a small cell (1-in. internal diameter) in which the sample is pressed at 1000 psi. The pressure is applied by a hand-operated screw, supplied with a calibrated spring prestressed so that the required stress is applied with a half turn of the screw. Electrodes consist of two circular rings inset into the bottom of the cell. The ground grain or flour is compressed against these coplanar electrodes, and as the current penetrates the sample only to a depth of the order of the electrode separation (0.04 in.) the quantity of sample in the cell is unimportant above a certain minimum. This dispenses with the need for weighing. The instrument appears to be well adapted for testing the moisture content of flour. With grain, the sample must first be ground. Brockelsby [36] claims that the loss of moisture is negligible during grinding in the small, enclosed, hand-driven mill provided with the instrument. Rough tests can be made on whole wheat when this is too wet to grind, say over 22%. Recent investigations made in the Grain Research Laboratory, Winnipeg, have established the standard error of estimate for wheat for this meter at about ±0.23%.

On grains the accuracy of resistance-type moisture meters has been found to be ±0.25% or better, which is the best available with standard oven-dry methods. For getting optimum operating conditions from the conductivity moisture meters the following factors should be taken into consideration: (1) moisture distribution, (2) range of measurement, (3) temperature, (4) sampling, (5) packing density, (6) purity of the sample, and (7) method of measurement. For detailed discussion of the effects of these factors, the reader is referred to Chap. V.

The three dielectric-type meters that appear most useful for testing cereals and their products are the Steinlite, the Motomco, and the SRI. To operate the Steinlite moisture meter, power is turned on and the circuit is balanced by setting the dial selector to a read button and turning a knob until the galvanometer reads 45 on the scale. A weighed sample (150 g of wheat) is put into a drop-bottom loading funnel. By tripping a release trigger the grain is introduced into a narrow cell of the meter, the long walls of which are plates of a parallel-plate condenser. To obtain the meter reading on the sample, the dial selector is rotated counterclockwise to select the correct radiofrequency range, and both the galvanometer reading and the button designation are recorded. Percent moisture is obtained from the instrument reading with the aid of appropriate tables. Comparison of an earlier model of the Steinlite meter with the vacuum oven gave an error of estimate of about ±0.4% moisture on hard red spring wheat.

The Motomco moisture meter (manufactured by Motomco Inc., USA) was originally developed by Rasmussen and Anderson [37]. The principle of operation of this meter is based on the balancing of two oscillating circuits as indicated by a milliammeter. One oscillator is crystal controlled and fixed to oscillate at 18 MHz, while the other, which contains the test cell for holding the grain, can be adjusted to balance with the fixed oscillator by means of a variable condenser coupled to a graduated dial. To make a moisture test with the meter, the temperature of the sample is taken, a portion is weighed (150 g of wheat for a 3-in. cell and 250 g for a $3\frac{1}{2}$-in. cell) and transferred to the dump cell and then released into the test cell. The meter circuit is balanced by adjusting a knob which is coupled to the dial. The dial reading is taken and this result together with the temperature of the sample is converted to percentage moisture from appropriate charts.

Hlynka et al. [35] have calibrated the Motomco electrical moisture meter (Model 919) for various grains; the standard error of estimate for samples tested ranged from 0.22% moisture for wheat to 0.39% for barley. The temperature correction required for wheat was found to be 0.045% moisture/1°F. The investigations undertaken were aimed at developing a practical moisture-measuring method for use by the Canadian Grain Inspection system. The electrical meter has been standardized and has replaced the Brown-Duvel method which takes a much longer time. It has been reported by Zeleny and Hunt [14] that the Motomco moisture meter was able to give consistently accurate moisture-content values when compared with standard oven-dry methods, whereas similar other electrical moisture meters showed variation and fluctuations in the values of the moisture content determination. The Motomco meter was found to be less influenced by electrical variations than other meters. It has also been less subject to area-to-area or season-to-season variations. Due to these reasons this moisture meter has been adopted for all grain inspection under the grain standards by Grain Division, Agricultural Marketing Service, United States Department of Agriculture.

Paull and Martens [38] have studied the factors affecting the determination of moisture in Canadian wheat by commercial electrical meters and found that the moisture values are affected by the size and shape of the wheat kernels but they are not influenced in any way by the protein or the ash content of the wheat samples. Differences in size and shape of the kernels can influence electrical readings in several ways. Variations in these properties result in variations in the closeness of packing in the dielectric-type meters. This in turn results in differences in the grain/air ratio with the resulting differences in the specific inductive capacity of the mixture [39]. The best thing to do for a uniform density attainment is that a constant weight of grain should be compressed to a constant volume for uniform results of moisture (content) measurement. Abnormal moisture distribution has been shown to have a marked effect on the electrical estimation of moisture (content). It has been observed by Harshorn and Mountfield [40] that an artificially damped sample tends to give unduly high readings. Abnormalities in the moisture distribution doubtless account also for the widely experienced difficulty in obtaining accurate meter estimates of moisture content in freshly harvested wheat and in wheat that has been rapidly dried. There is a high correlation between bushel weight and kernel weight so that it is difficult to conceive of the effects of these factors separately. Taken jointly they may be taken as good measures of kernel size and weight.

By incorporating a dielectric cell (cylindrical type) in the electronic unit of the SRI moisture meter, Pande [41] has measured the moisture content of a large number of samples of maize. The SRI moisture meter which is manufactured by Television and Electronics Associates works on the principle of the dielectric variations in accordance with the change in the moisture content of the sample (grains). During the course of investigation on the measurement of moisture content of the samples of maize it was found that the packing density of grains has an appreciable effect on the precision of the measurement. A grain dropper was therefore developed which was completely filled to the brim and the kernels of grain were dropped from its collapsible bottom at a fast speed resulting in a uniform packing of the sample in the dielectric cell. This arrangement gives a uniform packing density as shown by the consistency of the results obtained by this moisture meter which showed excellent agreement with the results obtained by the standard oven-drying method. The accuracy of this moisture meter was found to be ±0.25% in the range of 0 to 20% of moisture content. The temperature correction was found to be 0.05% in the temperature range of 60 to 100°F. The typical results obtained by this method are shown in Table 11.2.

Before concluding the description of the capacitance-type moisture meter, it is worthwhile to mention an important development that has taken place recently in the design of this type of meter which has improved the overall performance, range, and accuracy. This has been achieved by using a fringe capacitance by Thomas [42] who found that there is a linear relationship between moisture content and the fringe capacitance in the range of 0 to 10% whereas there is a logarithmic relationship between the moisture content (in the

TABLE 11.2

Comparative Values of Moisture (Content) of Corn as Determined by Electronic Meter in the Dielectric Cell and Oven-Dry Method in the Laboratory

Sample no.	Electronic meter reading (%)	Oven-dry values (%)
1	11.80	11.71
2	12.00	12.20
3	12.1	12.30
4	12.5	12.70
5	10.70	10.50
6	10.40	10.30
7	12.40	12.14
8	12.80	12.80
9	12.10	11.95
10	12.25	12.29

range of 5 to 45%) and the fringe capacitance. A fringe field probe has also been developed by the Foxbro [43] moisture-measuring system. This fringe field probe consists of an array of condenser electrodes embedded in a good dielectric with thin guard electrodes sandwiched in between. The polarity and potential of the measuring electrodes and the guard electrodes are arranged so that within the dielectric material no field exists between the measuring electrodes; all capacitance between measuring electrodes is in the air gap at the surface of the probe.

In order to get best results from capacitance-type meters, the following factors should be taken into consideration: (1) moisture distribution, (2) presence of electrolytes, (3) packing density, and (4) temperature effect. For detailed discussion of the effects of these factors, the reader is referred to Chap. V.

Readings on all electrical meters are also affected by the temperature of the grain. Accordingly the temperature of the grain must be known. Unfortunately, it is difficult to measure temperature of a sample of grain accurately. Any serious error in measuring the temperature of the sample leads to an erroneous estimate of moisture content. The United States Department of Agriculture recommends that the grain temperature should be allowed to come within $2°F$ of room temperature before tests are made.

Accurate calibration of electrical moisture meters presents certain difficulties. The best procedure appears to be that of using a large number of samples covering a wide range of natural moisture contents. A constant-temperature room in which all samples and the meter can be brought to uniform temperature is almost essential. The regression equation, established as a result of an adequately replicated study of this kind, should prove relatively accurate. Moreover, the data will show the moisture level at which the error of estimate begins to increase substantially. For most meters, increasingly less accurate estimates are obtained as the moisture content rises above 20%. The practice of calibrating meters with a few samples, artificially tempered or dried to give an adequate range of moisture levels, does not appear to be satisfactory.

Since bound water is not measurable by conductance-type meters, these meters measure free water only and add a constant value for the bound water. Charts for conductance-type meters are based on the assumption that there is a fixed amount of bound water at all times. However, there is evidence that the ratio of free and bound water may not remain constant, and as changes occur, errors are introduced in the moisture measurement. Although dielectric moisture meters measure both free and bound water, the great difference in dielectric value of these two types of water may introduce errors in measurement when the ratio of the two shifts. As mentioned earlier, however, the Motomco meter has consistently shown less deviation from the oven method values than any other dielectric meter tested. Electric grain moisture meters are, and will no doubt continue to be, widely used because of their great convenience, regardless of the fact that they are at times somewhat inaccurate.

2.1.7 Infrared Spectrophotometric Method. This technique has been extended to the measurement of moisture content of seed by methanol extraction. Recent work has shown that valid absorption spectra can be obtained on denselight scattering samples [45], so effort has been directed toward making quantitative measurements of the water absorption bands in solids. Water absorption bands occur at 0.76, 0.97, 1.18, 1.45, and 1.94 µm as well as at longer wavelengths [46-47]. The spectral region from 0.7 to 2.4 µm has been investigated by Norris and Hart [44] for measuring the moisture content of grain seeds. Two different instruments were used by these investigators. Preliminary measurements in the 0.7- to 1.1-µm region were made, using a special recording spectrophotometer designed for biological samples. Spectral absorbance curves were recorded for the samples under study and data were obtained from these curves to compute the difference in optical density between 0.97 and 0.90 µm.

The results of studies by Norris and Hart [44] with experimental instrumentation demonstrate the possibility of direct spectrophotometric determinations of moisture contents of grains and seeds and flour with little sample preparation. At 1.94 μm the sensitivity to moisture is very high. A sample thickness of 1.5 mm gives an optical density change of 0.04 for a 1% change in moisture content at normal levels. At very low moisture levels, this sensitivity increases to 0.10 for a 1% change in moisture. The measurement of optical density changes as low as 0.001 can be made readily with present photometric technology, so a moisture sensitivity of 0.05% should be possible. The limitation of the measurement is therefore the sample preparation. Duplicate samples gave variations in readings equivalent to as high as 0.5% in moisture content. This variation is from a combination of sampling error, sample preparation error, and instrumental error.

It should be possible to measure the moisture content of a wide range of materials by the use of the absorption band at 1.94 μm. The main requirement is for a uniform sample from 1 to 3-mm thick with no interfering absorption bands. Measurements on individual intact seeds should also be possible. For moisture contents greater than 20% the 0.97-μm band should give greater accuracy because the absorption at 1.94 μm becomes so strong that it is difficult to measure.

2.1.8 NMR Method. Nuclear magnetic resonance apparatus specially for the measurement of moisture content in solids such as cereals, grains, and other food products has been developed and manufactured commercially. One of the important commercial instruments is the Schlumberger NMR analyzer. An NMR apparatus (Fig. 11.5) for continuous measurement of moisture content of food products has been developed by Rollwitz [48]. Sample containers of a variety of sizes and shapes may be used ranging from 40 cm^3 down. No sample preparation is ordinarily required; the material can be placed in its original state in the container.

740								XI. MOISTURE IN FOODS

FIG. 11.5.   NMR apparatus for continuous measurement of moisture content.

When moisture measurements are required on flowing materials, techniques have been developed by Rollwitz [48] for passing the whole stream through the radiofrequency coil and the magnet as shown in Fig. 11.5. If the system is completely stabilized, a continuous measure of moisture content is obtained. The decreasing linewidth with increasing moisture content would make the peak-to-peak first harmonic amplitude-vs-moisture curve nonlinear rather than a straight line. Rollwitz [49] has developed a device which uses the variation in the linewidth with moisture level instead of the variation in amplitude. Therefore the system does not require consistent packing or density linewidth. This system is fully automatic and has an accuracy of 1%. It has been estimated from measurements that the accuracy can be maintained for filling factors of from 0.3 to 1.0, hence large voids do not cause errors.

This method has been successfully employed for the moisture measurement in hygroscopic materials. The resonance consists of a narrow line due to protons in the sorbed water superimposed on a broad line (20-40 kHz wide) due to protons in the solid absorbant. For lower moisture content (less than 10%) a small increase in the width of the narrow line is observed, but the change is usually of the same order as due to the linewidth of inhomogeneities in the field of the permanent magnet. Low moisture (content) measurements can thus be made by measuring the linewidth of the water signal. By using a high-resolution NMR apparatus, it has been possible to differentiate between the free and bound water. The time required for one analysis is 20 to 60 sec. Most of the instruments presently in use seem to be suitable for many types of hygroscopic materials over a wide moisture range (5 to 100%) and have an accuracy of 0.2%. Since hydrogen (proton) has the highest sensitivity of any nuclear species to detection by NMR, this technique appears to offer a rapid, nondestructive method of estimating moisture contents over a very wide range of moisture values with great accuracy. The particle size and packing in the case of granular solids are stated to have no effect on the signal absorption which is a great advantage over the electrical methods of moisture measurement, but it is essential to have a constant and correct weight of the sample. The sample should, however, be ground to small particle size for better results. Applications of NMR to moisture determinations in solids are virtually unlimited. These include wheat, oats, rice, sugar, starch and its derivatives, candy, corn, cheeses, and all types of cereals.

A continuous record of the moisture content of a sample of millomaize having different moisture levels is shown in Fig. 11.6 from which it will be observed that by using the NMR method, it is possible not only to measure the moisture content of an individual sample but the moisture content of the constituent units of the mixture having different moisture levels. This is a definite advantage and improvement over other methods of moisture measurement. By using the NMR method the moisture content of samples of starch has been measured with an accuracy of 1% or better. However, when fat is present in unknown amounts in such a sample it may cause a wide scatter in the data and it becomes difficult to establish a correlation. As the nuclear magnetic resonances are not specific for the moisture (H_2O) itself, but are specific for the hydrogen nuclei,

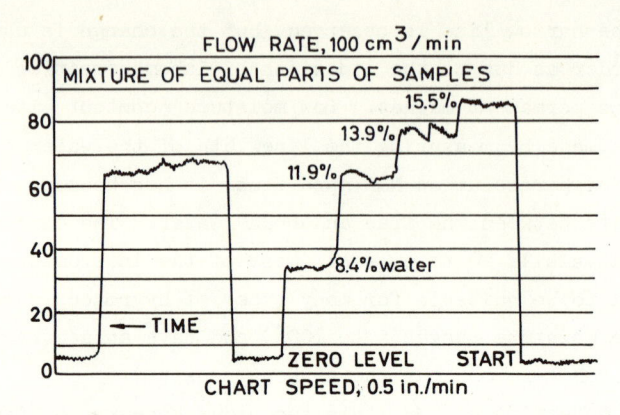

FIG. 11.6. Recording of the NMR peaks corresponding to different moisture contents of millomaize.

these measurements represent a function of all the hydrogen atoms present in the sample. If the hydrogen exists in two states, host molecules and water molecules such as starch and moisture, then the NMR signals from the two are generally easily separated. Thus NMR measurements of moisture in starch and like materials are readily made with accuracy over a wide range of moisture levels. At the very low moisture levels, however, the two signals may be difficult to resolve. Difficulty also exists for a material with two hydrogen-containing liquids adsorbed or absorbed. The NMR signals from these two liquids such as water and fat may be difficult to separate if a rapid determination is desired. If a little time is allowed, the two quantities may be resolved by drying, freezing, or adding paramagnetic ions and then using the saturation phenomenon. Quantitative measurements can be effectively performed on flowing materials. If there is a useful curve of linewidth as a function of moisture, then measurements can be made which are independent of density changes and voids.

3 MOISTURE IN FLOUR AND FLOUR PRODUCTS

3.1 Sorption of Moisture

A number of researchers [50-55] have studied the changes in weight and moisture content of stored flour and found that a composite sample of flour exposed in a still saturated atmosphere at a temperature of 23°C reaches a maximum moisture content of 28.74% in 9 to 12 days, at which time it becomes moldy. In a saturated atmosphere at 0°C a moisture content of 34.78% was obtained in 17 days which was not the maximum.

In Fig. 11.7 are shown the percentage of moisture in the patent and clear flours exposed to atmospheres of approximately 30, 50, 70, and 80% relative humidities. While the differences between the two grades of flour are small, they are in the direction of a slightly higher hygroscopicity on the part of the patent grade. Stockholm [54] had previously shown a difference in the hygroscopic moisture of starches prepared from patent and clear flours, those from the patent containing more moisture when they were exposed under identical conditions. The sorption curves are of the shape of a simple parabola, which, if extrapolated to 100% humidity, would give values in terms of hygroscopic moisture not far different from those reported for the flours exposed in a saturated atmosphere.

3.2 Methods of Measurement

The determination of the moisture content of flour presents complex problems due to the fact that after heating the starch present in the flour becomes hygroscopic and absorbs moisture from the atmosphere and makes the determination by ordinary oven-dry methods rather misleading. Due to this reason a number of new techniques and methods have been developed.

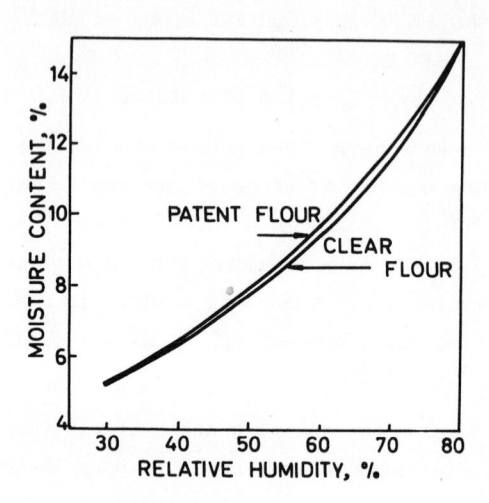

FIG. 11.7. Relation between moisture content and relative humidity in patent flour and clear flour.

A comparative study of oven-drying methods for the measurement of moisture in flours has been made by Smith and Mitchell [56]. The experimental work conducted by these researchers covers the determination of the moisture content of flour when employing three types of drying ovens in general use, viz., the water-jacketed vacuum oven, the water oven, and the electric oven, the variation that may be obtained in check determinations of the moisture in the same oven, the loss or gain in weight which flour may undergo when samples are exposed to the atmospheric conditions of the laboratory, both before and after drying, and the effect of cooling the dried samples under different conditions before making the final weighing for the determination of moisture.

When employing the official vacuum oven method for the determination of moisture in flour, Mitchell and Alfend [57] noted that greater percentages of loss in weight occurred if the samples were dried in loosely covered dishes than when dried in open dishes. Their work, however, did not show any comparison of results obtained on the same sample of flour when drying in covered and open dishes in other types of ovens in general use. For the purpose of making this comparison, the moisture on six samples of flour was determined. The results obtained show that the greatest loss in weight occurred when drying in the vacuum oven, the closely covered dishes giving higher losses and more uniform results than the open dishes. When drying flour is loosely covered with dishes at atmospheric pressure, the results are erratic. Theoretically, the covers would retard the evaporation of the moisture at atmospheric pressure; actually, this was found to be true.

Snyder and Sullivan [18] have determined the moisture content of flour by employing the hydrogen drying method; 58 flour moisture tests obtained by drying in a current of hydrogen yielded 0.54% more moisture than vacuum oven drying at 100°C and 600 mm pressure. When flours are dried in hydrogen, special precautions are necessary in order to secure a uniform flow of very dry gas over and through the flour masses.

Winton's hydrogen drying apparatus (Fig. 11.8) employed by these investigators with certain modifications yielded closely agreeing duplicate results; when nitrogen and preheated air were substituted for the hydrogen as the drying medium, practically the same results were obtained as with hydrogen. As the use of hydrogen is dangerous, it is desirable to use nitrogen and preheated air, which give the same results. These tests emphasize the importance of the dryness of the medium in which flours are dried in making moisture tests by any drying method.

Success of hydrogen, nitrogen, or air drying of flour in glass tubes placed within copper tubes heated to the temperature of boiling water depends mainly upon a uniform rate of flow of perfectly drying gas or air, and the loading of the tubes in such a way that the drying medium readily passes over through the flour. While the method is not suitable to routine work, where a large number of flour moisture tests are made daily, it is, however, a satisfactory research method, and gives when hydrogen is used (as here described) about 0.50% more moisture than vacuum oven drying at 99 to 100°C and 500 mm of vacuum. The method yields closely agreeing duplicate

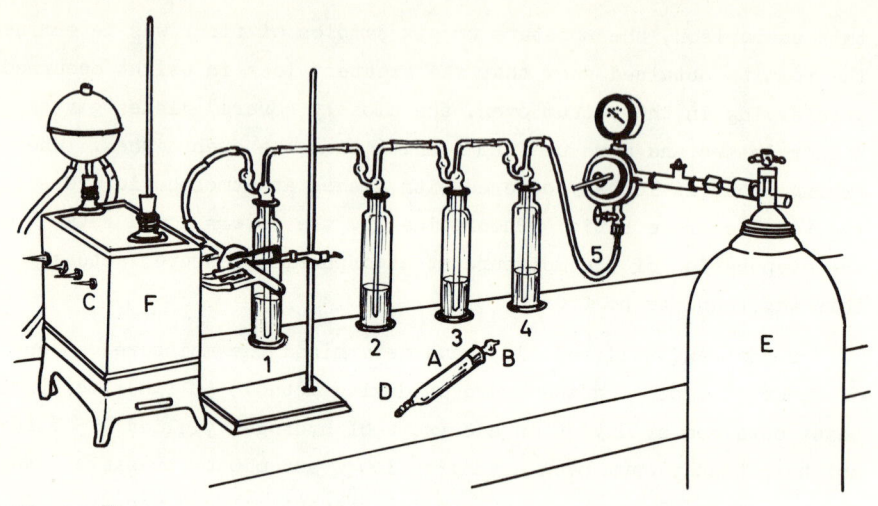

FIG. 11.8. Hydrogen drying apparatus.

results. Spencer [58], however, obtained lower moisture results with hydrogen drying than with any other method of drying.

It should be mentioned that as an industrial problem the determination of moisture in flour for manufacturing or other control purposes is largely a matter of using and strictly adhering to empirical conditions as to temperature, time, and manipulation. Such methods are not necessarily suitable for research work, because flour is a complex mechanical mixture of various carbohydrates, proteins, and other compounds, which are appreciably affected by heat. Hence any moisture method based on heating is a relative rather than absolute expression as to moisture content. By the term total moisture or water in flour is meant that which is held in loose and firm physical combinations with the flour particles, also that held in a condition closely approaching chemical combination with the starch micellae, or as water of hydration of proteins. It has been reported by Duval [59] that starch begins to lose water of hydration before all the moisture can be eliminated. He has applied a thermogravimetric method of analysis for the accurate determination of moisture content of flours in grams. The results obtained were checked by the Karl Fischer method and showed excellent agreement.

Hughes et al. [60] have developed a method for measuring moisture content of flour by using the Honeywell portable relative humidity instrument and placing the sensing element in the flour in order to attain rapid equilibration. It was found by these researchers that flour samples of widely varying moisture and protein contents conform rather closely to a single isotherm of relative humidity vs moisture content. As is well known, the Honeywell hygrometer is essentially a null-point resistance bridge with a temperature-compensating rheostat. The sensor is a Dunmore-type element containing lithium chloride.

The variation of equilibrium humidity with moisture content of two samples of flours is shown in Fig. 11.7. As will be seen from the curves they lie close to a single line on the plot indicating that the isotherms for the two flours are very similar. By this method the moisture content of flour can be determined rapidly and accurately, the accuracy being approximately 0.15% moisture for all types of flours and 0.05% if just one type of flour is used. This method was applied for the measurement of moisture content of starch as well with success.

Studies of the rate and magnitude of the moisture losses occurring during the drying of macaroni products involve a series of determinations at relatively frequent intervals extending over a period of several days. In this type of work it is necessary either to remove samples for the determination of residual moisture or to weigh directly a sample suspended within the drying cabinet. The removal of material at frequent intervals disturbs the drying schedule, introduces sampling errors, and is subject to the possibility of changes in moisture content during the operations involved. In view of these difficulties, direct weighing of a sample suspended from a balance external to the cabinet is preferable. Since the actual drying is carried out in a stream of conditioned air, as described by Bennington and Geddes [61], a certain amount of motion is imparted to the sample which renders it essential to use a type of balance insensitive to this condition.

In order to overcome the inconvenience of frequent weighings at regular intervals over extended periods and to secure a continuous record of the weight loss, an automatic recording balance was developed by Bennington and Geddes [61]. This automatic balance is electrically operated, is continuous recording, is fitted with oil damping, and is counterbalanced for an initial load of 100 g. Losses in weight up to 17 g with an accuracy of the order of 0.01 g are recorded by this balance without manual attention by means of an automatic device which places weights in the form of steel ball bearings upon the pan. Recording is performed, through the use of a timed spark, on a paper chart carried on the drum of a variable-speed hymograph which has a range of from 12 hr to 6 days. This automatic recording balance has been found to be a good control device in the manufacture of macaroni.

Moisture content of bread is determined either on the whole bread or on the crust and crumb separately. In the former case, the loaf is cut into four parts by two cuts at right angles. One part is weighed and then cut into thin slices which, together with any crumbs produced during cutting, are dried in a tared glass vessel in an air oven at 105 to 110°C for 7 to 8 hr, cooled in a desiccator, and weighed rapidly on a rough balance sensitive to 0.5 cg. The loss in weight represents moisture. If the percentage of moisture present when the sample is taken is required, the whole loaf should be weighed at that time and just before analysis; due allowance should be made in the calculation for the moisture lost in the meantime.

4 MOISTURE IN SUGAR AND SUGAR PRODUCTS

4.1 Sorption of Moisture

Moisture plays an important role in the storage and preservation of sugar and sugar products. It also has a marked effect on the composition properties of sugar and similar materials. The chief problem in keeping sugar in storage is to retain the free-flowing condition of the product. Problems of deterioration (inversion) and color development are also present, but of lesser concern under reasonable care. Lumpiness and deterioration can be caused by a moisture condition, and a knowledge of the behavior of sugar exposed to different atmospheric conditions is, therefore, essential. In more recent years, the literature has provided specific data, notably on the critical humidity value at which it is believed pure sucrose crystals change from a theoretically moisture-free condition to a dissolved or liquid condition. Whittier and Could [62] observed the humidity over a saturated solution to be 77.4% and Dittmor [63] presents graphical data placing the humidity value at about 83%. Rendon [64] gives 66.2% relative humidity as the safe maximum for storage of raw sugars; this indicates a safety factor of 0.265. It is apparent from various data that the purity of the samples and the existing temperatures play a part in setting the precise humidity value for storage.

There have been many studies on this subject, one of which by Keller [65] contains a useful review of the literature on deterioration in storage. Keller's studies showed no absorption of moisture at 50% rh, and appreciable absorption at 100%. The amount of moisture absorbed increases with increase of temperature at the higher humidities, e.g., at 30°C there is more than twice the absorption there is at 25°C with 100% humidity. Behne [66] suggests that critical relative humidity is 65% for normal ambient conditions. At this relative humidity, sugar will neither absorb moisture nor dry out. Webster [67] studied moisture absorption at 50, 62, and

78% humidity, respectively, each at 30, 35, and 40°C. The influence of ash, colloids, and reducing sugars together with polarization was also investigated. In general, all the sugars exhibit similar behavior, the moisture content increasing with both humidity and temperature, but the changes are not of the same order. Reducing sugar has a very large influence on the hygroscopic properties of sugars. Scott [68] found that raw sugar could be kept without change of weight for test for 9 months in storage.

The moisture-absorbing power of various sugars, polysaccharides, and sugar-containing products in an anhydrous condition and changes in their compositions have been investigated by Browne [69]. It was found that sugars whose safety factor exceeded 0.31 deteriorated fast, while those whose safety factors were below 0.29 suffered no appreciable change in composition. It was further observed that fructose-containing materials such as honey, molasses, and inert sugar are exceedingly hygroscopic at high atmospheric humidity, and therefore the sampling of such materials and the measurement of moisture content should be done at a much faster rate in order to prevent changes in moisture content and composition as a result of moisture absorption. The literature of moisture sorption in various materials is voluminous and the reader is referred to some recent collected publications [70-73] on the subject.

Observations made by McGinnis [72] involving the exposure of small samples of sugar at various known relative humidities between 30 and 80%, at room temperature, yielded results which are plotted in Fig. 11.9. It will be noted that the values suggest a sigmoid-shaped curve, with practically no moisture at low humidities, and that there is a rapid change in the range between 60 and 80% humidity. The trace of nonsucrose in the granulated sugars is a component that tends to absorb moisture gradually, and the sucrose theoretically would show an abrupt change. Thus the moisture content of the composite commercial product is in a sense an average of that of the two constituents. It may be noted that the differences in the nature of these nonsucrose impurities and in the amounts present

4   MOISTURE IN SUGAR AND SUGAR PRODUCTS   751

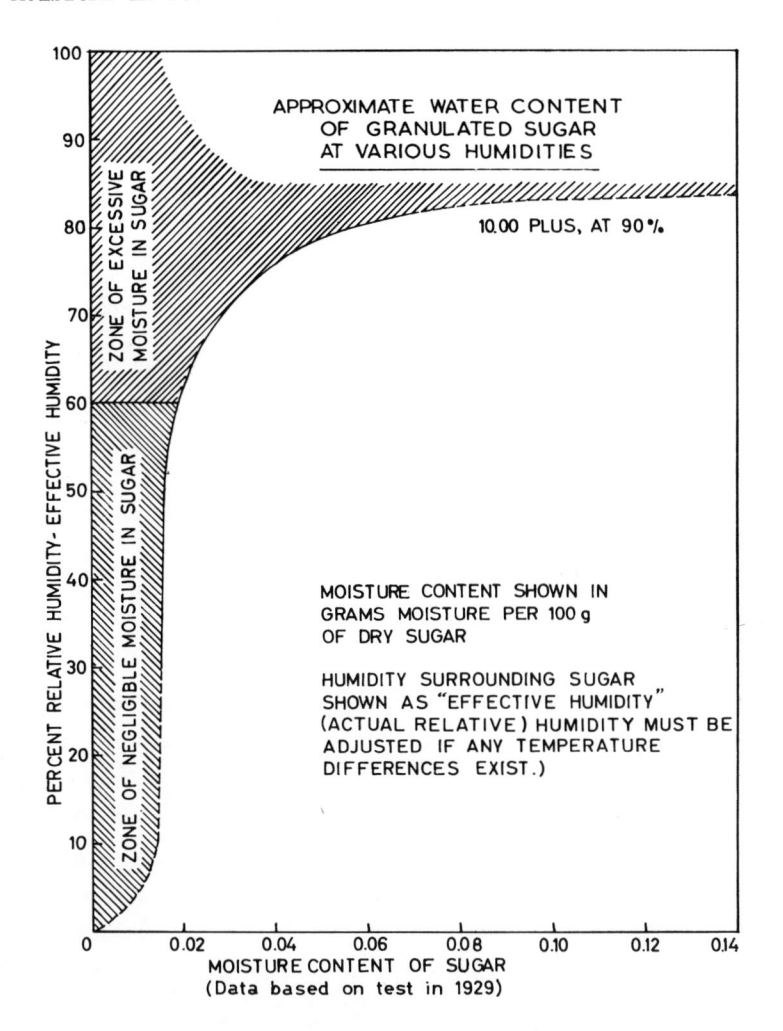

FIG. 11.9. Approximate moisture content of granulated sugar at various relative humidities.

will influence the humidity equilibrium values over the whole range and in particular will affect the slope of the curve or abruptness of the moisture increases in the critical storage range between 60 and 80% humidity.

4.2 Methods of Measurement

The problem of obtaining the correct moisture content of various saccharine products is as important to the sugar-laboratory chemist as is the knowledge of how to obtain correct data on the percentage of sugar or other constituents. There are two methods for the moisture determination of liquid and semiliquid saccharine products. These are drying on quartz sand according to the Association of Official Analytical Chemists [73] (USA); and the Official German method. These two methods are briefly described.

The procedure for drying on quartz sand is as follows: Digest pure quartz sand with strong hydrochloric acid, wash, dry, ignite, and preserve in a stoppered bottle. Place 6 to 7 g of the prepared sand, a short stirring rod in a flat-bottomed dish, dry thoroughly, cool in a desiccator, and weigh. Add 3 to 4 g of molasses, mix with the sand (if necessary), dry in a water oven at the temperature of boiling water for 8 to 10 hr, stirring at intervals of an hour, cool in a desiccator, and weigh. Stir, heat again for an hour, cool, and weigh, repeat the heating and weighing until the loss of water in an hour is not greater than 3 mg. Pellet [74] recommends that the material be mixed with pumice stone and dried at 102 to 105°C. It may, however, be mentioned that Gustavson and Pierce [75] have shown that sulfur dioxide, carbon dioxide, ammonia, iodoform-producing substances are given off by beet molasses when dried at 105°C.

The procedure for drying according to the Official German method is as follows: A ratio of 25 parts of iron-free sand to 1 part of dry substance is sufficient. Three grams of thick juice to 50 g of iron-free sand are weighed out. The drying takes place in vacuo at 105 to 110°C. For determining moisture content of massacuites, 2 to 3 g are intimately mixed with 50 g of iron-free sand in a moisture dish and given a preliminary drying for a quarter of an hour in a drying oven at 70°C. After again thoroughly mixing, the drying is continued for 6 to 8 hr at 105 to 110°C in a vacuum oven or air bath. The weight is taken as constant when after repeating the drying for a period of 2 hr the loss in weight is less than 0.10%.

Aikin [76] has determined the moisture content of beet sugar factory products and molasses by drying on sand; as molasses is one of the most impure products with which one has to deal, it was used as a basis for most of the work and the results were checked on purer products. It was assumed that the method that would give the highest results when using a temperature not above 110°C was the most nearly correct. All the drying process was done in an aluminum dish, 50 mm in diameter by 30 mm high, provided with a closely fitting cover.

The accurate determination of moisture, in some respects the most simple of analytical operations, is frequently one of the most difficult determinations that the sugar chemist is called upon to make. Among the chief difficulties that confront the chemist in determining the moisture content of sugar products by the ordinary methods of drying are as follows:

1. The very hygroscopic nature of many sugar-containing materials and the retention of water by absorption or occlusion

2. The extreme sensitiveness of some sugars, notably fructose, to decomposition at temperatures between 80 and 100°C with the splitting off of water and other volatile products

3. The liability of many impure sugar-containing substances on heating to give off various volatile products, such as alcohols, aldehydes, esters, organic acids, carbon dioxide, and ammonia, which are wrongly estimated as water in determining the loss of weight during drying. Oxidation may also occur in certain cases with formation of volatile decomposition products

The moisture determination is further complicated by the fact that many sugars such as maltose, lactose, and raffinose retain variable amounts of water of crystallization under different conditions of drying, so that the chemist is not always certain (even when no further loss of weight occurs in the oven) as to the exact amount of moisture that may be retained in the hydrated form.

In sugar-factory control, where determinations of moisture in raw sugars, bagasse, press cake, etc., are often desired with the greatest attainable speed, Spencer's electric oven [77] for rapid moisture tests is used to great advantage. This apparatus with cover removed and with cover in place is shown in Fig. 11.10. When the cover is in place, a rapid current of air, heated electrically by coils of resistance wire in the central cylinder H, is drawn by suction through perforated capsules F, which contain the product to be dried. The thermometer in the drying chamber is kept at the desired temperature (105°C for sugar, 130°C for bagasse) by means of a rheostat regulator. The moisture can be completely removed by this device from raw sugar in 10 min and from bagasse in 30 min without decomposition of product. A clock which operates an electric time switch and bell is placed in the circuit and indicates the completion of the drying period. Comparisons of determinations by Spencer's rapid electric oven with those obtained by the slower methods of drying show a close agreement of results.

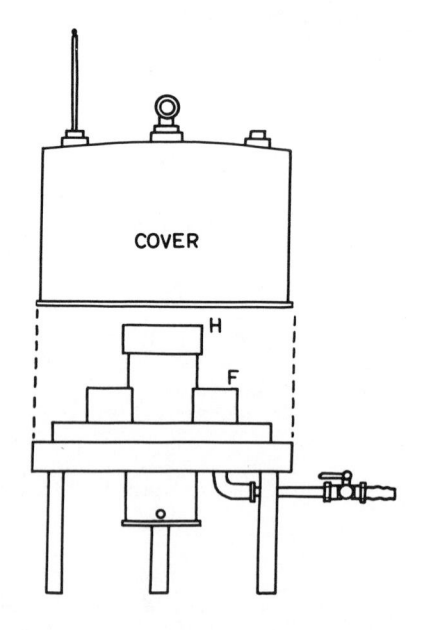

FIG. 11.10. Spencer's electric drying oven.

For dehydrating sirups, molasses, massecuites, and other sugar-containing substances, which contain but little or no fructose, the method of drying previously described may be used. The material, however, should first be absorbed on dry sand, pumice stone, or asbestos in order to facilitate the removal of the large excess of water. Methods of the Association of Official Analytical Chemists [73] (USA) which consist of drying on pumice stone and quartz sand are recommended for drying the semiliquid products of this class.

In a method of drying employed in France by Pellet [78], nickel capsules, 85 mm wide and 20 mm deep, are used. The capsule has a circular depression in the center. Each capsule is provided with a cover having a small notch at the edge for the passage of a small stirring rod. The raised border of the capsule is filled with fine particles (about 1 mm diameter) or freshly ignited pumice stone, employing an inverted funnel. The funnel is then removed, the cover and stirring rod put in place, and the capsule weighed. Then 3 g of the substance to be dried is weighed in the central depression of the capsule, 5 ml of hot distilled water is then added, and, after stirring to dissolve all soluble matter, the capsule is slightly inclined on different sides to permit absorption of the solution by the pumice stone. The process is repeated with 3 ml more of hot water and then with 2 ml. The contents of the capsule are then spread evenly over the entire bottom and dried in any suitable oven at a final temperature of 102 to 105°C.

In Java and some other countries filter paper is used as absorbing material as proposed by Josse [79]. A strip of filter paper 1 cm wide and 100 cm long is folded back and forth every 2 to 3 mm in zigzag fashion. It is then placed on a flat strip of the same width, the two strips are rolled up together, and the ends fastened with a pin. The large surface of this "rosette," with numerous air spaces, facilitates the removal of water vapor. The rosette is first dried to constant weight at 102 to 105°C in a weighing dish. It is then removed from the dish, 2 to 3 g of molasses or sirup is weighed into the dish and diluted with about 5 ml of water. The rosette is replaced in the dish and quickly absorbs the liquid, which is then dried to a constant weight in the oven at 102 to 105°C.

Owing to the susceptibility of fructose to decomposition in the presence of water at temperatures much above 70°C, the methods previously described are not applicable to the determination of moisture in such products as honey, sugar-cane molasses, jams, fruit products, and similar substances. The error which may result from this source has been estimated experimentally by Carr and Sanborn [80] by hydrating a solution containing 17.75% of fructose. The solution was dried on pumice stone in flat-bottomed dishes at 100°C in air. It was observed that the percentage of solids after 6 hr drying is lower than the actual amount of fructose taken.

The susceptibility of many sugar products to decomposition at 100°C in air induced many early researchers to propose drying in vacuum. According to Weisberg [81] and Carr and Sanborn [80], the dehydration at low temperatures under reduced atmospheric pressure is the only recognized method for the accurate determination of moisture in fructose-containing materials. A number of vacuum drying methods and techniques have been developed for moisture determination in sugars, molasses, and sirups. The reader is referred to the original publications [80-81] for experimental details.

A few distillation methods have been applied successfully for the measurement of moisture content of sugar products. Testoni [82] determined the water in molasses by measuring the quantity that was obtained on distilling the product with turpentine of bp 160°F. Later, Van der Linden, Kauffman, and Leistra [83] devised a direct method for determining the water in molasses and other sugar-factory products by distilling 50 g of the product with 350 ml of xylene in a copper distillation flask connected to an upright condenser, the lower end of which discharged into a 250-ml measuring cylinder graduated to 0.05 ml.

On account of the liability of certain sugars, more especially fructose, to decompose at the boiling point of xylene (bp 139°F),

Bidwell and Sterling [84] have modified the distillation method by employing toluene (bp 110.5°) as the immiscible liquid with which the product is boiled. The correction for loss of water by volatilization and absorption was determined by Bidwell and Sterling [84] by making a blank distillation of 75 ml toluene with an accurately measured amount of water. It was found to be +0.02 ml. A comparison of water determinations of various sugar-containing products by the toluene-distillation method and vacuum-oven method gave good agreement.

Cleland and Fetzler [85] have developed the technique of the filter cell method for determination of moisture content of sugar products. This method gives better results. Zerban and Sattler [86] have applied the Karl Fischer method for moisture determinations of sugar products. Almy et al. [87] have employed the Karl Fischer method for the measurement of moisture content of a large number of products containing monohydric and polyhydric alcohols as well as of mixtures ranging from thick sirups to thin process solutions.

Based on the refractive index measurement a new method for estimating moisture content of honey has been developed by Chataway [88]. For absolute measurements, the refractive index of a substance is referred to vacuum. Chataway [88] has determined a relationship between moisture content and refractive index. The refractive index of the honey sample is measured by the Abbe refractometer and moisture content is determined. Temperature corrections at the rate of 0.00023°C should be applied if the measurement is done at a temperature other than 25°C.

An electrical moisture meter based on the dielectric properties of moist sugar has been found to be the most promising for measuring the moisture content of raw sugar in the laboratory by Trott and Barrow [89] and Trott and Saint [90]. These investigators have made use of a commercial moisture meter known as the Kappa moisture meter. However, any other moisture meter based on the same principle of working can be equally applicable for the measurement of the moisture content of raw sugar provided it is accurately calibrated as has been done in the case of the Kappa moisture meter.

The Kappa moisture meter was also used [89-90] as a control instrument for controlling the moisture content of the sugar leaving the centrifuge in the sugar factory. This was done by taking a continuous sample from the conveyor between the centrifuges and the storage bin and testing the accumulated sample with the meter at a fixed time interval. The meter was also used with considerable success by Trott and Saint [90] in checking the output from an automatic centrifuge at an early stage in the curing of each strike and this allowed prompt and positive adjustments to be made to process timing and control equipment. An additional advantage for factory use is that the instrument can be sited at almost any position in the factory and the only ancillary equipment required is a balance of up to 2 1/2-kg capacity sensitive to about 1 g at full weight loading.

Mekkawi [91] has used the Kappa moisture meter for measuring the moisture content of raw sugar and found that the empty cell reading of the meter was noticed to fluctuate between 7 and 11.5 units. He further observed that the relation between the relative humidity of the atmosphere and the meter readings was not proportional.

Munro and Wise [92] have employed the Kappa dielectric moisture meter for the measurement of moisture content of diluted molasses. The meter reading is considerably influenced by the presence of other substances in molasses; as the molasses concentration is increased, the meter reading increases at first very rapidly, becoming slow later on. By taking readings on pure solutions it was observed that the effect is associated with the electrolytes rather than with the sugars in the molasses. This suggests that in considering the use of the dielectric meter on sugar products, the ash content might be of some importance, and this point should be borne in mind in investigations with bagasse and sugar.

5 MOISTURE IN DRIED FRUITS

5.1 Introduction

Dried fruit is frequently stored in bulk over a period of several months, during which it is possible for considerable changes of moisture content to occur. These have an important effect on the appearance and keeping quality of the fruit. Conditions in various countries produce an increase of moisture content, and under extreme conditions this may result in the growth of molds on the fruit, which is thus rendered unsalable. The moisture content is also an important factor in the fumigation of dried fruit with ethylene oxide. The higher the moisture content, the greater is the absorption of fumigant. The chief factors influencing the direction of the change are the original moisture content [93] and the humidity and temperature [94] of the atmosphere. Under practical storage conditions, the amount of the change is dependent on the method of packing of the fruit, the method of storage in piles, circulation of air, and the ambient temperature and relative humidity.

Storage life of dried fruits is closely related to their moisture content. It was generally concluded by the early investigators that browning in dried-out fruits is decreased by decreasing the moisture content. Jewell [95] emphasized the importance of maintaining low moisture levels in the storage of dried apricots. Miegand et al. [96] recommend low moisture levels for the storage of all cut fruits. It has been a general observation in industry that dried fruits placed in bin storage darken more rapidly if the moisture content is high (20 to 25%). Nichols and Reed [97] stated that "drying to approximately 10% moisture improved the colour permanence of apples; apricots were unchanged; pears were poorer for the treatment." It should be emphasized, however, that in spite of this general conclusion that high moisture content is detrimental, there is very little experimental evidence in support of the literature.

Culpeper and Cladwell [98] studied the influence of humidity on the storage quality of dried apples. They suspended the fruit in muslin bags in storage chambers (bell jars of 11,000-ml capacity) at relative humidities of 0, 8.5, 18.8, 37.1, 47.7, 58.3, 70.4, 88.8, and 100% and stored them at 25°C (77°F). Higher humidities favored a rapid rate of darkening. Samples at 0% humidity remained in good condition for over 3 years, at which time they contained 0.4 to 0.7% moisture. Stadtman et al. [99] have shown that the influence of moisture on darkening of dried apricots is greatly modified by the presence or absence of oxygen. Under anaerobic conditions a maximum rate of darkening was found to occur between 5 and 10% moisture. Increasing the moisture content from 10 to 25% caused a 15 to 30% increase in the storage life of apricots at 37 and 49°C in an oxygen-free atmosphere. As the fruit was exposed to increasing quantities of oxygen, however, the beneficial effect of high moisture became progressively smaller, and in the presence of very large amounts of oxygen (200 mg O_2/100 g of fruit) the rate of deterioration was nearly the same for fruit at all moisture levels. The detrimental effect of high moisture when fruit is stored in the presence of large amounts of oxygen was shown to be due to a greater rate of oxygen uptake [100]. These results emphasize the importance of carefully controlling oxygen in studies on the influence of moisture in relation to the browning of apricots. Differences in the tightness of pack (ratio of oxygen to fruit) must be avoided. The interdependence of oxygen and moisture in the deterioration of dried fruit other than apricots does not appear to have been studied. It may be that, as with apricots, the detrimental effect of moisture on browning so often observed is actually due to an increase in the rate of oxygen uptake.

According to Wright et al. [101] a higher relative humidity is desired for storing fresh fruits and vegetables and florist stocks than for cereal grains. Leafy vegetables are normally stored slightly above freezing at 90 to 95% relative humidity; bulb plants

5 MOISTURE IN DRIED FRUITS

like onions, garlic, and gladioli at 70 to 75%, and fruits at 85 to 90% relative humidity.

5.2 Moisture Sorption

Experiments were conducted by Brown [102] for the determination of equilibrium conditions required for the storage of dry fruit. The determinations were carried out at 18 and 23°C in a relative humidity in a constant-temperature room. The relative humidity of the air in the cabinet with an accuracy of 1% was determined [103] by a dew-point hygrometer. The time taken to reach equilibrium between the air and the fruit varies with the moisture content of the fruit. When this is 12% or more, equilibrium is attained within 3 hr, but if the moisture content is about 8% the time taken is 20 hr or more.

The results are shown graphically in Fig. 11.11. It is at once evident that temperature has only a small effect on the equilibrium values. As would be expected for a sample of given moisture content an increase of temperature decreases the equilibrium values of relative humidity. The curves cover the range of relative humidity of 45 to 75%. A moisture content of 7 to 8% corresponding to the lower end of the range is below the value normally found for dried fruit. A moisture content of 23 or 24% attained in an atmosphere with a relative humidity of about 75% is also rare.

Two values were also obtained [102] for currants. These were substantially the same as would be obtained with sultanas, although in each case the value lay a little above the curve. Thus at 23°C currants with a moisture content of 13.1% were in equilibrium with an atmosphere of 56% relative humidity, and at 18°C currants of 15% moisture content were in equilibrium with an atmosphere of 61.2% relative humidity.

The problem of moisture change occurring during bulk storage of dried fruits has been extensively studied by Brown [102] on a large scale by using samples of 60 tons of dried fruit in the vaults

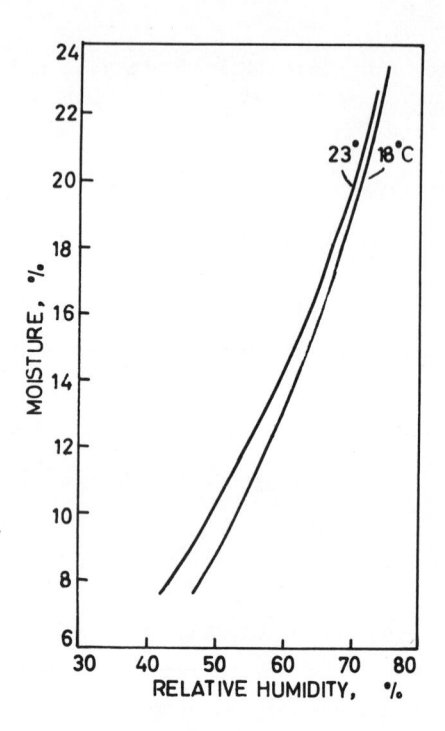

FIG. 11.11. Sorption isotherms of dried fruits.

of a London dock. It was found that the maximum moisture content attained was 22.6%. It was further observed that the surface layer of fruit in boxes exposed to the humid atmosphere showed considerable increase in moisture content and consequently greater signs of decay. However, the rate of increase was found to be slower for fruits in the interior layers, where a maximum moisture content of 7.5% could only be attained after a storage period of 28 weeks.

5.3 Methods of Moisture Measurement

An inexpensive form of vacuum oven (Fig. 11.12) was used by Brown [102] for this purpose, the sample being dried for 6 hr at 70°C ± 1°C. Determinations are made in copper boats 15 cm long and semicircular in cross section, the diameter being 3.5 cm. The number of each boat was stamped on the ends. A single lid is provided

5 MOISTURE IN DRIED FRUITS

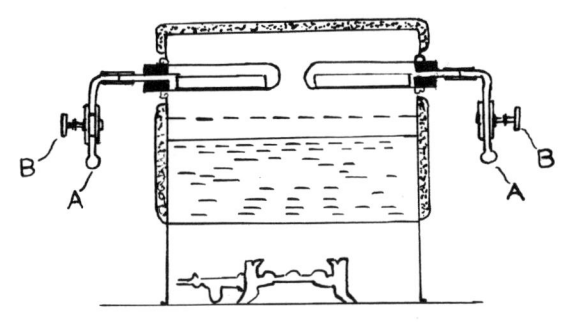

FIG. 11.12. Brown vacuum oven.

which fits all the boats used. Each boat slides inside a thick-walled glass tube, closed at one end and carrying at the other end a rubber bung fitted with a short glass tube and a piece of rubber pressure tubing.

The temperature of 70°C is recognized as the most suitable temperature for drying products such as dried fruit containing a large proportion of fructose which decomposes at temperatures in excess of this [103]. A series of determinations made by Brown [102] with different periods of drying showed that most of the water was expelled in 6 hr, only a further 2.5% being removed in another 6 hr. Moreover 6 hr is a convenient working period. In a series of determinations made over several days on the same sample of fruit, there was little increase of accuracy when the drying period was extended to 8 hr instead of 6.

The electrical conductance or resistance method has also been used for moisture measurement of dried fruits. The electrical moisture meter developed by Sykes and Coote [104] has been applied for dried apples. The efficiency of a meter to provide accurate estimates of the moisture content of dried apple is assessed by comparing the values obtained from calibration tables of moisture and electrical resistance with those obtained by the accepted laboratory method developed by the Association of Official Analytical Chemists [105]. This procedure is adopted with a Calipco meter [106] and also with an instrument of modified design, now known as the

Stowell meter, designed and constructed at the CSIRO Regional Laboratory [104] (Australia). The Stowell meter is smaller in size and faster to operate under factory conditions than the original Calipco meter. The point of balance is indicated on a microammeter instead of by the audiosignal used in the Calipco meter. The scale of the ammeter mounted on the panel ranges from 500 µA at one end, through zero to 500 µA at the other. A resistance and fine adjustment switch protect the ammeter should the resistance become grossly out of balance. The cell and electrode assembly are constructed [104] as a separate unit which is connected to the main instrument when in use and clamped conveniently into the lid during transport. Calibration tables for converting meter readings into moisture content have been prepared for the Stowell meter, the electrical circuit of which is shown in Fig. 11.13. The moisture content is obtained directly from the calibration tables. In determining moisture content by the Association of Official Analytical Chemists [105] (AOAC) method, minced material is dried at 70°C for 6 hr under a vacuum of at least 26 in. Hg. Dry air is admitted at the rate of at least two bubbles per second. The data used in the study were obtained by making meter measurements and AOAC determinations on a large number of samples of dried apples obtained from several factories and made from a number of varieties of apples at varying maturities.

The values of moisture content as obtained on laboratory and commercial samples of dried apples by using the Calipco electrical moisture meter and the AOAC method are illustrated graphically in Fig. 11.14. The data were analyzed by Sykes and Coote [104] on the assumption that for the subsamples tested the AOAC method gave a true value with a high degree of precision. The data obtained by the Stowell moisture meter using commercial samples showed a greater degree of precision as compared with the AOAC method. Also this meter showed greater accuracy and precision as compared with the Calipco meter. The standard error of estimates for single determinations on single samples were about ±0.80 and ±0.40 from the Calipco and Stowell meters, respectively. The reason for the highly

5   MOISTURE IN DRIED FRUITS

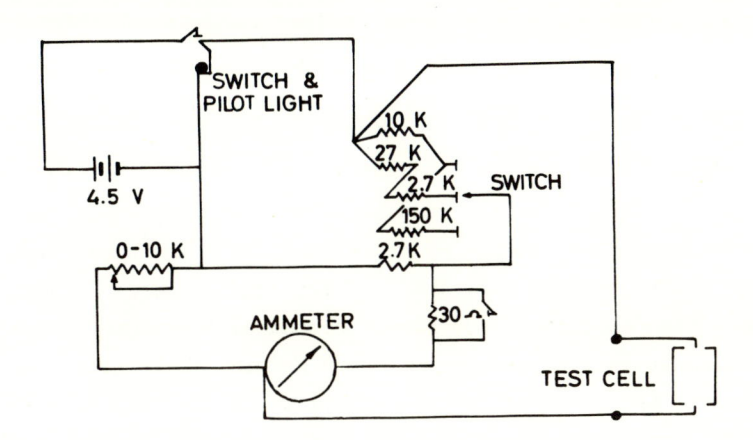

FIG. 11.13. Electrical circuits of the Stowell moisture meter.

significant difference between commercial and laboratory samples as shown in Fig. 11.14 was not investigated by Sykes and Coote [104]. It is probable that the SO_3 ion content, which would certainly have been greater in laboratory-prepared samples, was sufficient to result in a change in conductivity.

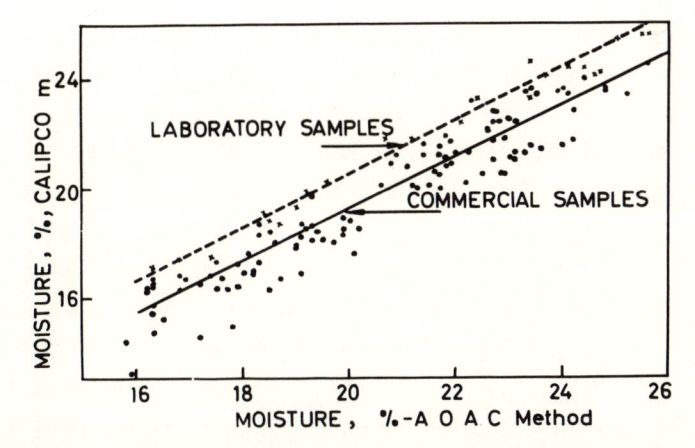

FIG. 11.14. Comparative values of moisture content of laboratory and commercial samples of dried apples obtained by the Calipco meter and the AOAC method.

6 MOISTURE IN DEHYDRATED FOODS

6.1 Introduction

Dehydrated foods can be classified into four groups, each group being characterized by a different sorption characteristic. The group comprises (1) starchy, (2) proteinous, (3) high sugar—high molecular weight constituent, and (4) high sugar foods. The moisture content of the dehydrated food has a tremendous influence on its storage life as well as flavor and other keeping qualities. The importance of the level of relative humidity to the stability and keeping quality has been the subject of a great deal of investigation; quite an appreciable amount of work has been done on this subject and this branch of science forms an important contribution in this field of food technology.

The workers of an Industrial Research Laboratory of Continental Can Company (USA) [107] conducted storage tests on commercially dehydrated potatoes. In most instances, the samples were acceptable after 1 year at room temperature, but darkened considerably and became inedible after 6 months at 36.7°C (98°F) and quite dark after 1 month at 64.4°C (130°F).

Howard [108] determined the effect of absolute moisture content on the storage life of diced potatoes containing about 300 ppm of sulfur dioxide and packed in nitrogen. The relation of absolute moisture content to storage life has been determined by the method of Makower et al. [17]. This method differs from the usual vacuum oven drying method in that the entire sample is ground to 40 mesh and a portion of the ground sample is held in a vacuum oven at 70°C (158°F) for a longer period of time (40 hr). They also found that the higher the moisture content, the more rapid was the development of the brown color. Caldwell [109] noted that potato strips discolored more rapidly at high temperatures than at low ones and the samples stored in air with a high relative humidity discolored more rapidly than those stored in dry air. They claimed that browning practically ceases when the moisture content of the material is reduced below 5%. Data were obtained also on the relation of moisture content to the storage of dried potatoes. Moisture levels of approximately 1% were obtained by packaging with calcium oxide in accordance with the method described by Howard [108]. Storage life was almost doubled by a decrease in moisture of 2.5%. It was found that in practically all cases a straight line was obtained when the data were plotted on semilog coordinates, using percentage moisture as abscissas and storage life in days as ordinates. The slope of the line was nearly always the same, apparently being independent of variety, and of reducing sugar and sulfite content. The observations made with dehydrated potatoes apply equally well to other similar vegetables and fruits.

6.2 Moisture Sorption

Salvin [110] classified dehydrated foods into four groups, each group being characterized by a different shape of isotherm. The groups comprised (1) starchy, (2) protein, (3) high sugar—high molecular weight constituent, and (4) high sugar foods. In addition, Salvin [110] indicated the importance of the level of relative humidity to the oxidative stability of dehydrated foods. He postulated that a certain amount of water in the form of a monomolecular layer on active molecular sites may be protective against oxidation in certain foods. Practical application of these concepts, as well as interpretation of the deteriorative and protective relationships in water sorption, necessitates more information on the effect of controlled humidity conditions on storage stability of dehydrated foods. According to Giles [111] the types of adsorption isotherms obtained for dehydrated foods vary with their sugar content. Those containing much starch or proteins give Brunauer type II curves but those containing high proportions of sugar give Brunauer type III curves. Intermediate types occur with foods having moderate sugar contents.

The equilibrium moisture content of some dehydrated vegetables was determined by Makower and Dehority [112] by allowing them to attain equilibrium in air-free desiccators containing sulfuric acid solutions to control the relative humidity. Fresh vegetables were used for desorption measurements and dried vegetables for adsorption. Experiments were carried out on carrots, cabbage, yams, spinach, and white potatoes. The sorption isotherms for all the vegetables are S-shaped and are characterized by an inflection point in the neighborhood of 5% moisture content. Measurements were also made on blanched white potatoes. Blanching causes a decrease in the equilibrium moisture content. The decrease is ascribed to a change in the physical state of the starch granules in the potato. The isotherms for various vegetables obtained by these investigators are shown in Fig. 11.15. These isotherms provide information for the limiting conditions of the dehydration process. Regardless of whether vacuum or air drying is used, the isotherms make it possible to specify the maximum permissible pressure or the maximum permissible humidity to reduce the moisture content to a desired value. Knowledge of equilibrium moisture content is also important in the study of packaging of dehydrated vegetables. The equilibrium values yield information concerning the vapor pressure of water within the package and are thus helpful in estimating from known permeability data the effectiveness of the packaging membrane in protecting the dried material against moisture uptake.

The effect of moisture contents, relative humidities on shelf life, and keeping qualities of some dehydrated compounded soup powders have been investigated by Rao [113]. It was observed by Rao that in onion soup powder the critical moisture content was 4.5% corresponding to 40% rh while in the other soup powders the respective values were 6.0 to 6.5%. The lower critical moisture content in the case of onion soup powder was obviously due to the extremely hygroscopic nature of dehydrated onions. Above these moisture levels the free-flowing characteristics of the soup powders were adversely affected and they tended to form lumps. At equilibrium relative humidities ranging from 84 to 93%, mold growth was visible after about 11 days in split bean soup, 18 days in cauliflower,

6 MOISTURE IN DEHYDRATED FOODS

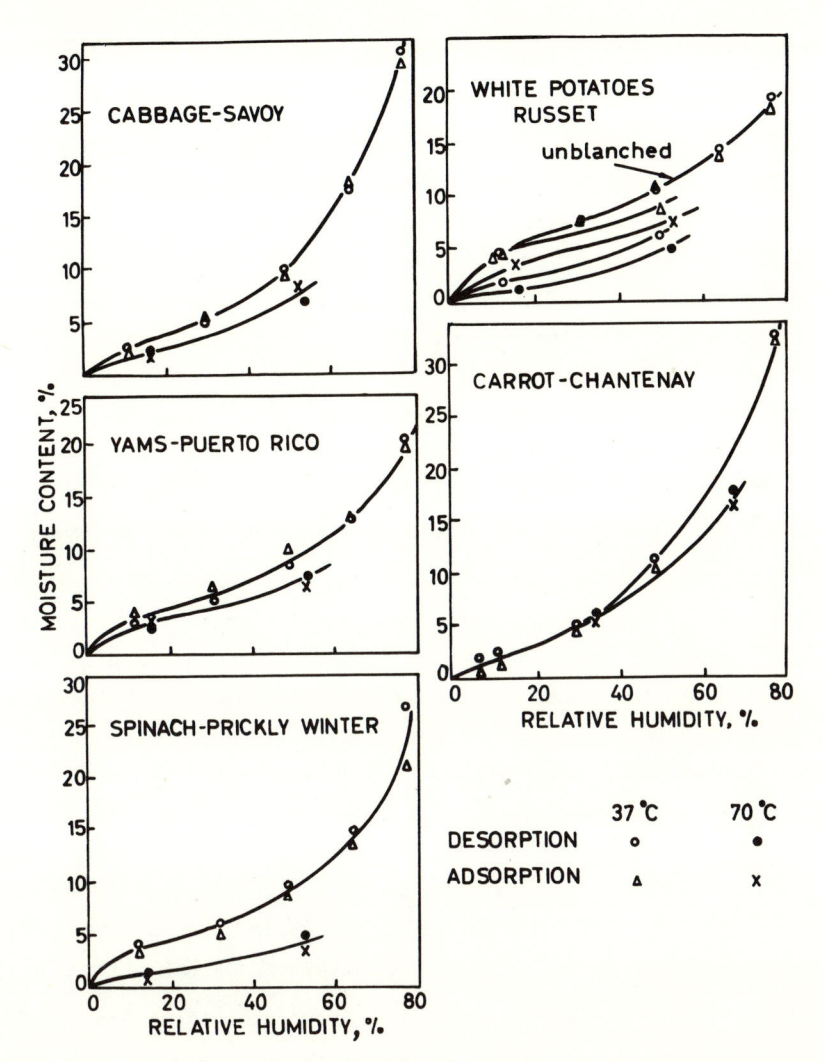

FIG. 11.15. Absorption-desorption isotherms of dehydrated vegetables.

mixed vegetable and chicken soups, and 31 days in onion soup powders. Sulfur compounds in onions seem to delay the appearance of mold in the onion soup powder.

6.3 Methods of Measurement

Makower and Nielsen [114] have developed an interesting reference method for determination of moisture content of dehydrated vegetables involving addition of a large amount of water to a weighed sample of vegetable, freezing and drying in the frozen state, and completion of the drying in a vacuum oven or vacuum desiccator in the presence of an efficient water absorbent. The last step can be completed in a relatively short time, at or slightly above room temperature, because of a marked increase in drying rate brought about by lyophilization. Data presented by these researchers for white and sweet potatoes, beets, and carrots show that the lyophilized materials can be dried unambiguously to constant weight and that the loss in weight may be taken as a measure of the moisture content.

A number of methods and techniques for the determination of the moisture content of egg powders have been developed and applied with varying degrees of success. These are (1) the AOAC oven method, (2) the fast oven-drying method, (3) the distillation method, and (4) the Karl Fischer titration method. The Fischer titration method has been reported by Frediani et al. [115] to give excellent results on egg powder. In a search for a more reliable method for determining moisture in dehydrated food materials having low residual moisture content (below 5%), the Karl Fischer titrimetric procedure has been investigated over a period of years by Shroeder and Nair [116].

Johnson [117] has applied the Fischer volumetric method for the measurement of moisture content of dry fruit materials such as wheat starch, potato starch, sucrose, glucose, fructose, pectin, spray-dried whole eggs, and a number of dehydrated vegetables and fruit powders.

Shivadjain [118] has developed a hygrophotographic technique for measuring moisture content of chemical and biological materials which are rather difficult to measure by most of the methods and techniques described so far. Hygrophotography is the name given to the process by which an image can be obtained on a sensitive surface by combined action of light and humidity. The principle which is new and unknown for other chemical compounds is based on the reversibility of a photochemical reaction accompanied by a change of coloration which disappears instantly under the action of a very small quantity of water and of humidity. It therefore permits the recording of the water emitted by a source of humidity at the place of its emission, and in this way it associates chemical sensibility with photographic precision.

This technique has been employed by Shivadjain [118] to estimate the degree of hydration of sausages and other porks in the following manner. A slice of sausage 2 to 3 mm thick is pressed between two hydrophotographic plates and after the preparation has been placed on a table, a suitable pressure is exerted on the upper plate until the appearance of the hygrophotographic image is sufficiently detailed. The two plates are then removed and copies are made either on transparencies or on paper. To operate more rapidly, a copy is made on transparencies, and a negative of the transparencies is prepared on an identical plate which serves in turn to make prints on photographic paper or enlargements.

7 MOISTURE IN MILK PRODUCTS

7.1 Introduction

Dried milk is produced either by spraying a fine mist of liquid milk into a heated chamber (spray dried) or by drying a thin film of the liquid milk on a hot roll (roller dried). A product containing approximately 4% water results. A greater water content than this is conducive to spoilage by microorganisms, whereas a very low water content is conducive to fat oxidation. In addition to liquid whole milk, the following milk products are dried: skim milk, buttermilk, whey, and cream. Dried milk is widely used in various foods.

Butter is basically a perfect emulsion. About 15% by volume is the aqueous, dispersed phase; the remainder is a semisolid, fatty phase. While nearly all the moisture is dispersed in small droplets, there also exist small undispersed inclusions or continuous channels. It has been shown by Prentice [119] that the incidence of these inclusions has considerable bearing on the keeping quality of the butter. The total moisture content is also important economically, particularly in view of legal requirements in many countries. The two phases in butter have dissimilar electrical properties. This difference has been utilized in studying the moisture distribution.

7.2 Methods of Measurement

Although moisture in dried milks may be obtained by drying to constant weight at 100°C in a vacuum oven, the toluene distillation method is more generally used since the high sugar content of dried milk causes some difficulty when drying at elevated temperatures.

The conductivity of butter provides an assessment of the undispersed moisture, if certain precautions are taken in sampling and in making the measurements. Dielectric theory predicts a dependence of the permittivity of an emulsion on the volume fraction of the dispersed phase. With butter there is considerable divergence between theory and experiment. This divergence may be due to distortion of the droplets or to an uneven distribution of electrolyte among the droplets. It is suggested that both causes may operate simultaneously. King [120] has reported that water in dehydrated butterfat is more accurately and quickly determined by Karl Fischer titration than by the oven method.

Moisture dispersion and distribution in butter has been determined by Herman and Lomward [121] by making electrical resistance measurement of a column of butter. Resistance was measured using a Wheatstone bridge and tubular electrodes of stainless steel, the longer tube being filled with butter and the smaller tube embedded on it. The resistance provided a reliable indication of moisture content in well-worked butter. Poorly worked butter shows considerable variation with the sample. Salt has a slight effect on resistance. Changes in the dispersion of moisture in the butter during working are described in the publication of these investigators.

The hygrophotographic method [118] has also been applied for the measurement of moisture content of butter and margarine. When a hygrophotographic plate is exposed to light and is ready for use, a section of a uniform and freshly made lump of butter or any other greasy solid material, such as margarine, having dimensions in proportion with those of the chosen plate, is then applied to the gelatin of the plate. The lump is held in place for 4 to 5 min if butter is used, and longer for a product less rich in water; during this time a light pressure is exerted on it. When the hygrophotographic image appears on the side of the support, the mass of greasy material is then taken off and the plate is cleaned carefully with a piece of cotton in order to eliminate the last portion of fatty material sticking to it. Then a print is made by contact process on silver bromide paper or on transparencies, and the picture obtained of the distribution of the humidity in the greasy material is permanently attached.

Continuous on-line measurements of the moisture content of butter have been attained by Gönczy and Nadai [122] who developed an instrument based on the measuring-alternation principle. The monitoring instrument was fitted in a continuous butter-churning machine in a dairy plant in 1970 and it has been functioning continuously ever since with great accuracy.

The microwave attenuation method has been employed by Kraszewski [123] for the continuous measurement of water content of margarine. For calibration, a correlation between the microwave meter readings and water contents as determined by standard gravimetric methods was established. An average difference between these two values was obtained which was less than ±0.25% H_2O over the range 14.5 to 18.5% H_2O.

8 MOISTURE IN RAW BEVERAGES

8.1 Introduction

The moisture content of raw coffee is an important characteristic, the exact and precise knowledge of which interests all those who are handling this product. The effect of moisture content and relative humidity on the storage of coffee beans has been studied by Natrajan et al. [124]. They determined the equilibrium moisture content values of coffee beans at different relative humidities.

Majumdar et al. [125] have studied the role of moisture content in the storage of coffee beans under airtight conditions. It was found by them that green coffee loses its color very rapidly if the initial moisture content at the time of storage is high. In the preliminary experiments, it was observed that a moisture content up to 10% was suitable for storage. For these trials, therefore, well-dried coffee having a moisture content of 9 to 10% was selected from the curing works. Beyond a moisture content of 10.5%, the color of plantation coffee fades with temperature and time of storage. The results of the experiments on the storage of plantation coffee in various containers have indicated that there is no moisture uptake in the sample stored in a stoppered glass bottle and polyethylene. However, it was noted that 200-gauge polyethylene did not give protection against moisture to the desired extent. At low humidity, coffee samples stored in gunny bags recorded a loss of chlorogenic acid, while in all the other containers stored both at low and high humidity there was no significant change. Majumdar et al. [125] have also studied the effect of moisture content and relative humidity on the bulk storage of coffee beans in bins. It was observed by them that the initial moisture content had a critical effect on the color of plantation coffee when stored under hermetic conditions. There was a significant change in color in coffee with an initial moisture of 12.8% during storage for 3 months at room temperature in sealed glass bottles. Under bulk storage conditions in an RCC bin, due to the wide temperature fluctuations between day and night hours, condensation of moisture was observed in the outdoor bin. In the outdoor bin the condensation of moisture was less, obviously due to less fluctuation in the diurnal temperature.

8.2 Methods of Measurement

Guilbot [126-127] was the pioneer who introduced a fundamental reference method of measurement of moisture content of raw coffee. He used coffee samples with normal water content of 8 to 11%. In the case of samples having moisture contents higher than 10 to 12%, preheating is done to reduce the moisture content to a value low enough to be compatible with good grinding. It consists in placing the sample in a drying chamber at 100 to 102°C for a period during which the water content comes up to the desired value. However, preheating should not be pushed too far, because then the residual moisture content falls to a rather low value and if dry grains thus obtained are ground very fine there is a chance of their absorbing moisture from the ambient atmosphere.

Although the fundamental reference method introduced by Guilbot [126-127] described above gave a standard determination of the moisture content in the low-moisture region (below 11%), it was not able to give correct and rapid results at moisture content values higher than 11%. To overcome this drawback, d'Ornano and Guilbot [128] developed other methods which were found applicable to coffee samples having moisture contents higher than 11%. These researchers also, employing the fundamental reference method, employed drying at a faster rate of heating the samples at 102°C for 15 hr and at 130°C for 5 hr. By heating raw coffee samples at different temperatures involving different durations they could obtain the rate-of-drying curves for the samples of raw coffee. They selected raw samples of coffee which contained moisture contents of 17.78%, 15.56% of which were placed in a drying chamber for 15, 30, 45, 60, and 90 min. In order to measure the loss of weight suffered by each of the samples, they were placed in a heating chamber (whose temperature was maintained at from 100 to 102°C) for 15, 30, 45, 60, and 90 min. By this method they measured the loss in weight suffered by the different samples which were cooled for 30 to 45 min in a desiccator and weighed on a precise balance. They obtained a set of curves (Fig. 11.16) from which it will be observed that no matter what the initial moisture content of the sample, the heating of the sample for 45 min in the drying chamber at 100 to 102°C brings the sample to a moisture content value of from 8 to 10%. From this it can be concluded that 45-min duration of heating is the right order of time corresponding to a temperature of 100 to 102°C at which the sample can be dried without any risk of any other volatile matter evaporating out of the raw coffee sample. A product is considered to be dried when it is in equilibrium with an atmosphere whose partial vapor pressure is zero.

These researchers took a sample of 5 g of raw coffee which was predried and ground properly. However, a more correct procedure of sampling is that 15 g of this sample should be taken predried, ground in bulk, and divided into three equal parts. This method of sampling is likely to eliminate the divergence of results due to heterogeneity of the lot in grains.

d'Ornano [129] has also employed an electrical moisture meter based on the variation of dielectric constant for the measurement of moisture content of raw coffee. The electrical moisture meter used is known as HYB 22. This moisture meter has been manufactured by Compagnie des Compteurs de Montronge, France, and consists of a stationary condenser in the form of a hopper (in which the grains of raw coffee whose moisture content is to be determined are placed) and a variable rotating condenser on the dial of which is attached a paper disk graduated directly in percentage of moisture content. Whenever there is a change in the capacity of the stationary condenser due to moisture content of the sample, the resonance of the circuit is disturbed, which is automatically restored by the rotating variable condenser, and the amount of movement suffered by the rotating condenser gives directly the value of the change in capacitance which gives the value of moisture content. There is also a

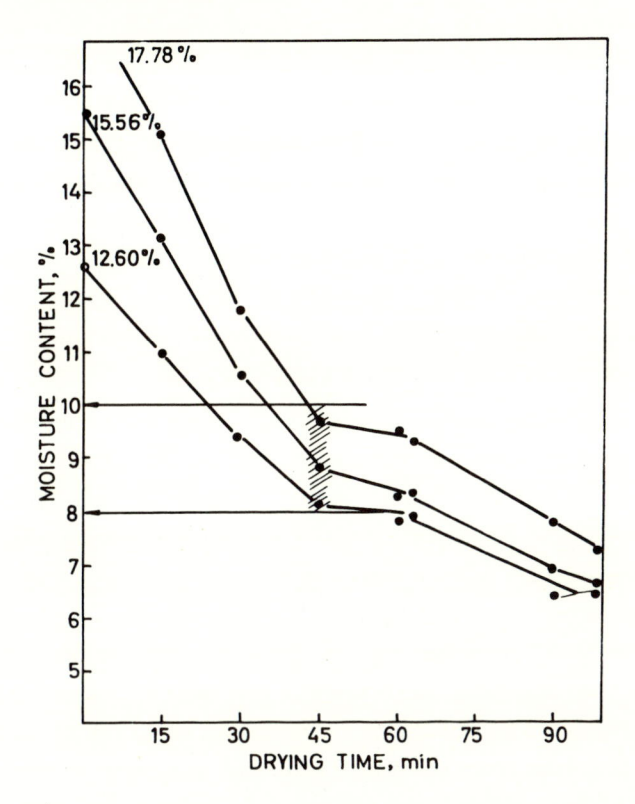

FIG. 11.16. Rate-of-drying curves of raw coffee samples.

provision for making corrections for the variation of temperature if the measurement is made at a temperature other than 20°C for which this moisture meter has been calibrated and standardized. The results obtained by this apparatus were found to be fairly accurate and reproducible; the standard deviation was found to be ±0.6%, and therefore for all commercial applications the moisture meter can be used with advantage.

Moisture content of cacao powders has been accurately measured by Lily Fozy [130] by using the dielectric method. It was found by her that the dielectric meter is suitable for moisture determination in other confectionary powders having similar particle size. She also obtained curves for temperature correction.

Tea, like coffee, is hygroscopic to a certain extent and the moisture sorption characteristics of this beverage follow the same pattern as that of coffee. Final tea coming out of the drier should have 3% moisture, and at the time of packing it would have from 4 to 5%, depending on climate. However, especially during wet weather, if the tea is not cured well it might even have 4.5% moisture while leaving the drier and further gain another 1 to 1.5% before packing, in which case a percentage of 6% could be found. If this happens, the keeping quality of tea drops considerably.

The usual oven-dry method using forced hot air to dry tea samples is generally used for the determination of moisture content. The temperature of the air for drying the samples of tea is usually regulated between 120 and 250°F. The duration of heating depends on the moisture content of the tea, and the temperatures used for drying the tea have to be standardized. For example, by trial and error it may be found that heating at 220°F for 20 min could be sufficient to find moisture percentage. The object is that no further loss of moisture should be observed in the last two readings, and at the same time the tea should not give off a burning smell.

Another method which is quite popular in tea estates for the determination of moisture percentage is the use of the infrared moisture meter. In this instrument infrared radiations are used for quickly drying the tea samples without appreciably raising their temperature. The instrument is a sensitive torsion wire instrument, the torque being applied to one end of the wire by means of a calibrated drum to balance the loss of weight as the test sample dries under infrared radiation. The drum is calibrated directly to read moisture percentage on the wet weight basis. The speed of drying (10 to 15 min for most materials) thus attained in combination with the frictionless balancing system gives results as accurate as those obtained by the standard oven-drying methods.

9 MOISTURE IN CONDIMENTS AND SPICES

The moisture content of spices and condiments has an effect on these materials similar to that on other food items discussed earlier as regards flavoring properties, storage, and other keeping properties such as shelf-life and packaging conditions. It is not possible to use any type of oven-drying technique for the measurement of moisture content of spices and condiments as the flavoring essence or principle present in these materials is most adversely affected by heating above 50°C. Therefore, recourse to other methods and techniques of moisture measurement has been necessary and toluene distillation is one of the methods which is very commonly used and has been described in Chap. III.

This method of moisture measurement can be made applicable to all sorts of spices and condiments. According to IS 1907-1961 for cardamon, the same procedure can be applied with equal success for this material. However, for the determination of moisture content of asafoetida, which is also used in India as a flavoring agent or a condiment, a vacuum-oven drying method has been used by Shastry et al. [131]. The values of the moisture content of asafoetida range from 3 to 6%. It may, however, be pointed out that the moisture content of raw asafoetida when it is collected in the form of "tears" is of the order of 30%, which after being dried attains a moisture content in the range from 3 to 6%. The estimation of the moisture content of asafoetida plays an important role in estimating the alcoholic extract of this condiment, which is a criterion of the quality of the asafoetida for commercial purposes.

10 MOISTURE IN TOBACCO

Moisture in tobacco is commonly determined by the statutory method as loss in weight on drying at a defined temperature. During recent years attention has been directed to the effect of variations in the time of drying and in the degree of ventilation in ovens;

such variations, it has been shown by Fryd and Koff [132], may seriously affect the apparent moisture content. The sensitivity of the apparent moisture content to drying conditions is demonstrated to arise, at least in part, in a component which does not exist as moisture in the undried tobacco but is produced by reactions of the Millard type or by other reactions involving reducing sugar during the process of drying. An apparatus has been described by these investigators by the use of which experimental variations in the apparent moisture content of tobacco can be minimized. As this apparatus is similar to the vacuum-drying oven, it is not described here.

According to Indian Standard Specifications No. 5643 of 1970, moisture in tobacco should be determined by drying the sample in an air-circulated electric oven at a temperature of 100°C for 15 hr.

REFERENCES

1. L. Zeleny and D. A. Coleman, *Cereal Chem.*, 15, 580-585 (1938).

2. W. V. Hukill, *Humidity and Moisture*, 2, 1965, p. 116.

3. J. D. Babbitt, *Nature*, 156, 265-266 (1945); *Can. J. Res.*, 27F, 55-72 (1949).

4. F. C. Fenton, *Agr. Eng.*, 22, 185-188 (1941).

5. J. E. Hubbard, F. R. Earle, and F. R. Senti, *Cereal Chem.*, 34, 422-33 (1957).

6. C. K. Shedd and H. H. Thompson, *Agr. Eng.*, 35, 786-788 (1954).

7. S. Brunauer, P. H. Emmett, and E. Teller, *J. Am. Chem. Soc.*, 60, 309-319 (1938).

8. M. Stephenson, *J. Text. Inst.*, 29, T297 (1938).

9. G. F. Davidson and S. A. Shorter, *J. Text. Inst.*, 21, T165 (1930).

10. J. H. Thompson, *Ind. Chem.*, 34, 451-53 (1958).

REFERENCES

11. S. G. Barker and J. J. Hedges, British Research Association for Woollen and Worsted Industries, Publication No. 71, 1926, p. 13.

12. J. G. Downes, B. H. Mackay, L. G. Bellamy, and V. D. Burgmann, Brit. J. Appl. Phys., 38, 484 (1961).

13. U.S. Department of Agriculture, Agricultural Marketing Service 1959, "Methods for Determining Moisture Content as Specified in the Official Grain Standards of the United States and in the United States Standards for Beans, Peas, Lentils and Rice," Service and Regulatory Announcements, No. 147 (Revised), p. 3.

14. L. Zeleny and W. H. Hunt, "Moisture Measurement in Grain," Paper presented at the 1962 Winter Meeting, American Society of Agricultural Engineers, Chicago, Ill., Dec. 1962.

15. Association of Official Analytical Chemists, "Official Methods of Analysis," published by the Association of Official Analytical Chemists, Benjamin Franklin Station, Washington, D.C., 8th ed., 1955.

16. R. B. Brock, Chem. Ind., 299-302 (1947).

17. B. Makower, S. Chastain, and E. Nielsen, Ind. Eng. Chem., 38, 7, 725 (1946).

18. H. Snyder and B. Sullivan, Part I, Ind. Eng. Chem., 16, 741-44 (1924); Part II, Ind. Eng. Chem., 18, 272 (1925); Part III, 17, 311-314 (1926).

19. D. A. Coleman and E. G. Boerner, "The Brown Duvel Moisture Tester and How to Operate It," U.S. Department of Agriculture, Bulletin 1375 (1926), revised 1936, Mimco, Supplement 1942.

20. K. Hlynka and J. A. Anderson, "The Brown Duvel Moisture Testing Method as Used by the Grain Inspection Branch," Board of Grain Commissioner for Canada, Grain Research Lab. Board of Grain Commissioners for Canada, Winnipeg, Canada, Mimco, Bulletin (1950).

21. E. Brown and J. W. T. Duvel, U.S. Bureau Plant Industry, Bull. 99, 1907.

22. D. A. Coleman and E. G. Boerner, U.S. Department of Agriculture, Dept. Bull. No. 1375 (Revised), 1936, 44 p.

23. L. Sair and W. R. Feitzer, Cereal Chem., 19, 714-720 (1942).

24. R. H. Fosnot and R. W. Hamon, Cereal Chem., 22, 41-49 (1945).

25. J. R. Hart and M. H. Neustadt, *Cereal Chem.*, **34**, 26-37 (1957).
26. E. L. Weise, R. W. Burks, and J. K. Taylor, *Natl. Bur. Std. Tech. News Bull.* 47, 1963, p. 116.
27. H. A. Daynes, *Gas Analysis by Means of Thermal Conductivity*, Cambridge Univ. Press, 1933.
28. E. B. Jones, *Instrument Technology*, vol. 2, *Analysis*, Butterworths, London, 1933.
29. D. H. Desty, *Vapor Phase Chromatography*, Butterworths, London, 1957, p. 75.
30. G. Zweig and J. Sherma, *Handbook of Chromatography*, The Chemical Rubber Co., 1972.
31. W. M. Schwecke and J. H. Nelson, *Anal. Chem.*, **36**, 689 (1964).
32. P. J. Geary, *Control*, **7**, 303-5 (1963).
33. G. S. Fraps and A. R. Kemmerer, *Texas Agr. Expt. Sta. Bull.*, 557 (1937).
34. G. S. Fraps and A. R. Kemmerer, *Ind. Eng. Chem. (Anal. Ed.)*, **13**, 806-9 (1941).
35. I. Hlynka, V. Martens, and J. A. Anderson, *Can. J. Res.*, **27F**, 382-97 (1949).
36. C. F. Brockelsby, *Cereal Chem.*, **28**, 83-94 (1951).
37. H. E. Rasmussen and J. A. Anderson, *Can. J. Res.*, **27**, 249-52 (1949).
38. A. E. Paull and V. Martens, *Can. J. Res.*, **27**, 479-82 (1949).
39. L. G. Grover and J. King, *J. Soc. Chem. Ind. (London)*, **63**, 320-24 (1945).
40. L. Harshorn and J. D. Mountfield, *Milling*, **10**, 24-7 (1944).
41. A. Pande, *Instrum. Pract.*, **15**, 432 (1961).
42. A. M. Thomas, *J. Sci. Instrum.*, **43**, 21 (1966).
43. H. W. Gebele, Moisture Measurements in Industry, *IRE Trans. Ind. Electron.*, **7** (1962).
44. K. H. Norris and J. R. Hart, *Humidity and Moisture*, vol. 4, Reinhold, New York, 1965, pp. 19-25.

REFERENCES

45. W. L. Butler, *J. Opt. Soc. Amer.*, <u>52</u>, 292 (1962).

46. J. F. Collins, *Phys. Rev.*, <u>26</u>, 771 (1925).

47. J. A. Curcio and C. C. Petty, *J. Opt. Soc. Amer.*, <u>41</u>, 302 (1951).

48. W. L. Rollwitz, Moisture Measurement in Various Hygroscopic Materials Using Nuclear Magnetic Resonance, Proceedings of the International Symposium on Humidity and Moisture held in Washington, D.C., 1963.

49. W. L. Rollwitz, U.S. Pat. 3,045,175 (1958).

50. W. H. Brewer, Relation of Grains to Moisture, *Tenth Census Reports*, <u>3</u>, 28 (1833).

51. F. B. Guthrie and G. W. Norris, Daily Variations in the Moisture Content of Flours, *Sci. Bull. (New South Wales), Australia*, <u>7</u>, 18 (1912).

52. T. Sanderson, *Dakota Sta. (USA), Spec. Bull.* <u>3</u>, 1914, p. 14.

53. C. O. Swanson, J. T. Willard, and L. A. Fitz, *Kansas Sta. (USA), Bull.* <u>202</u>, 1915, p. 119.

54. W. L. Stockholm, *Dakota Sta. (USA), Bull.* <u>120</u>, 1917, p. 1.

55. C. H. Bailey, *Ind. Eng. Chem.*, <u>12</u>, 1102 (1920).

56. E. R. Smith and L. C. Mitchell, *Ind. Eng. Chem.*, <u>17</u>, 180 (1925).

57. L. C. Mitchell and R. Alfend, *J. Official Anal. Chem.*, <u>8</u>, 76 (1924).

58. G. L. Spencer, *Ind. Eng. Chem.*, <u>13</u>, 70 (1921).

59. C. Duval, *Inorganic Thermogravimetric Analysis*, Elsevier, Amsterdam, London, 1963, p. 675.

60. P. J. Hughes, J. L. Vaala, and R. B. Koch, *Humidity and Moisture*, vol. 1, Reinhold, New York, 1965, p. 133.

61. D. S. Bennington and W. F. Geddes, *Ind. Eng. Chem., (Anal. Ed.)*, <u>8</u>, 76 (1936).

62. E. O. Whittier and S. P. Could, *Ind. Eng. Chem.*, <u>22</u>, 77 (1930).

63. J. H. Dittmor, *Ind. Eng. Chem.*, <u>27</u>, 333 (1935).

64. Q. D. Rendon, *Philippine Agr.*, <u>19</u>, 383 (1930).

65. A. G. Keller, *Sugar J.*, **2**, 25 (1939).

66. H. Behne, *Sugar J.*, **2**, 284 (1940).

67. J. Webster, *Intern. Sugar J.*, 46 (1941).

68. J. J. Scott, *Proc. Louisiana Sugar Cane Tech. Conf.*, 1939.

69. C. A. Browne, *Ind. Eng. Chem.*, 178 (1918).

70. A. W. Adamson, *Physical Chemistry of Surfaces*, Interscience, New York, 1960.

71. F. Daniels and R. A. Alberty, *Physical Chemistry*, Wiley, New York, 1955.

72. R. A. McGinnis, *Beet Sugar Technology*, Reinhold, New York, 1951, p. 387.

73. Association of Official Analytical Chemists, *Methods of Analysis, AOAC*, 5th ed., 1940, p. 484; also *J. Assoc. Official Anal. Chem.*, **23**, 88 (1940).

74. H. Pellet, *Pellet's Method of Determining Moisture, Physical and Chemical Methods of Sugar Analysis*, vol. 29 (C. A. Browne and F. W. Zerban, eds.), Wiley, 1941; *Fribourg's Anal. Chim.*, **9** (1907).

75. R. G. Gustavson and J. A. Pierce, *Ind. Eng. Chem.*, **16**, 167 (1924).

76. V. L. Aikin, *Ind. Eng. Chem.*, **12**, 979 (1920).

77. G. L. Spencer, *J. Assoc. Official Anal. Chem.*, **8**, 305 (1925).

78. H. Pellet, *Pellet's Method of Determining Moisture, Physical and Chemical Methods of Sugar Analysis*, vol. 29 (C. A. Browne and F. W. Zerban, eds.), Wiley, 1941; *Fribourg's Anal. Chim.*, 90-94 (1907).

79. A. Josse, *Bull. Assoc. Sucr. Dist.*, **10**, 656 (1892/93).

80. O. Carr and T. F. Sanborn, *Bull. No. 47, US Bureau of Chem.*, 134-151.

81. C. Weisberg, *Bull. Assoc. Chim. Sur. Dist.*, **11**, 524 (1893/94).

82. G. Testoni, *Staz. Sper., Agrar. Ital.*, **37**, 366 (1904).

83. Van der Linden, M. Kauffman, and F. Leistra, *Arch. Suikerind*, **25**, 951 (1917).

REFERENCES

84. G. O. Bidwell and W. F. Sterling, Methods of Analysis, AOAC, 5th ed, 1940, p. 353.

85. J. E. Cleland and W. R. Fetzler, Ind. Eng. Chem. (Anal. Ed.), 13, 858 (1941); 14, 124 (1942).

86. F. W. Zerban and L. Sattler, Ind. Eng. Chem. (Anal. Ed.), 18, 138 (1946).

87. E. G. Almy, W. E. Griffin, and C. S. Wilcox, Ind. Eng. Chem. (Anal. Ed.), 12, 392 (1940).

88. H. D. Chataway, Can. J. Res., 6, 532 (1932).

89. R. D. Trott and R. W. Barrow, Determination of Moisture Content of Raw Sugar by Means of Kappa Moisture Meter, Proc. 1960 Meeting of British West Indies Sugar Technologists, held at Jamaica, pp. 185-87.

90. R. R. Trott and J. Saint, A Rapid Method for the Determination of Moisture in Raw Sugars, Proc. 11th Congress of I.S.S.C.T., Mauritius, 1962, Elsevier, Amsterdam, 1963, pp. 959-65.

91. F. M. Mekkawi, Proc. 11th Congress of I.S.S.C.T., Mauritius, 1962, Elsevier, Amsterdam, 1963, p. 966.

92. R. E. C. Munro and W. S. Wise, Proc. 1960 Meeting of British West Indies Sugar Technologists, held at Jamaica, p. 176.

93. C. M. Johnson, Ind. Eng. Chem., 17, 312 (1945).

94. J. G. Kapsalis, M. Wolf, M. Driver, and A. S. Henick, The Moisture Sorption Isotherm as a Basis for the Study of Sorption and Stability Characteristics in Dehydrated Foods, Proc. 16th Research Conference of the Research Council of American Meat Institute Foundation of the University of Chicago, USA.

95. W. R. Jewell, J. Dept. Agr., Victoria, 35, 498 (1937).

96. E. H. Miegand, H. S. Madsen, and F. E. Price, Food Packer, 24, 604 (1943).

97. P. F. Nichols and H. M. Reed, Western Canner and Packer, 23, 11 (1931).

98. C. W. Culpeper and J. S. Cladwell, J. Agr. Res., 35, 889 (1927).

99. E. R. Stadtman, H. A. Barker, V. A. Hass, E. M. Mrak, and G. Mackinney, Ind. Eng. Chem., 38, 324 (1946).

100. E. R. Stadtman, H. A. Barker, V. A. Hass, E. M. Mrak, and G. Mackinney, Ind. Eng. Chem., 38, 99 (1946).

101. R. C. Wright, D. H. Rose, and H. Whiteman, The Commercial Storage of Fruits, Vegetables, and Florist and Nursery Stocks, U.S. Department of Agriculture Handbook No. 66, 1954, p. 178.

102. W. B. Brown, J. Soc. Chem. Ind., 57, 31 (1938).

103. J. Stanley, Appl. Biol., 23, 428 (1936).

104. S. M. Sykes and G. G. Coote, The Rapid Estimation of Moisture in Dried Apples, C.S.I.R.O. (Division of Food Preservation), Technical Paper No. 29, 1962.

105. Association of Official Analytical Chemists, Official Methods of Analysis, 8th ed., 1955, p. 343.

106. M. A. Joslyn, Methods of Food Analysis, Applied to Plant Products, Academic, New York, 1950, p. 71.

107. Research Staff, Continental Can Co., Foods Inds., 16, 991 (1944).

108. L. B. Howard, Canner, 100, 46 (1945).

109. J. S. Caldwell, Canner, 100, 35 (1945).

110. H. Salvin, Food Technol., 17, 34 (1963).

111. C. H. Giles, Chem. Ind., 724 (1964).

112. B. Makower and G. L. Dehority, Ind. Eng. Chem., 35, 193 (1943).

113. E. G. Rao, Defence Science J. (India), 14, 43 (1964).

114. B. Makower and E. Nielsen, Anal. Chem., 20, 856 (1948).

115. H. A. Frediani, J. T. Owen, and J. H. Bird, Trans. Amer. Soc. Cereal Chemists, 10, 176 (1952).

116. C. W. Shroeder and J. H. Nair, Anal. Chem., 20, 452 (1948).

117. C. M. Johnson, Ind. Eng. Chem., 17, 312 (1945).

118. J. Shivadjain, Humidity and Moisture, vol. 2, Reinhold, New York, 1965, p. 656.

119. J. H. Prentice, Humidity and Moisture, vol. 2, Reinhold, New York, 1965, p. 185.

REFERENCES

120. L. F. King, Austr. J. Dairy Technol., 17, 312 (1945).

121. M. N. Herman and S. H. Lomward, S.A.F.R. J. Agr. Sci., 5, 439 (1962).

122. J. Gönczy and B. Nadai, Proc. Imeko Symposium on Moisture Measurement, Hungary, 1971, pp. 325-40.

123. A. Kraszewski, Proc. Imeko Symposium on Moisture Measurement, Hungary, 1971, pp. 405-15.

124. C. P. Natrajan, S. K. Majumdar, K. S. Srinivasan, A. Balachandran, D. S. Bhatia, and V. Subrahmanyan, Food Sci., 10, 315 (1961).

125. S. K. Majumdar, M. Muthu, and J. K. Krishnarao, Food Sci., 10, 322 (1961).

126. A. Guilbot, Café Cacao, Paris, Janvier-Mars, 7, 49-56 (1963).

127. A. Guilbot, Café Cacao, Paris, 7, 192-200 (1963).

128. M. d'Ornano and A. Guilbot, Café Cacao, Paris, 8, 293-300 (1964).

129. M. d'Ornano, Café Cacao, Paris, 8, 113-132 (1964).

130. Lily Fozy, Proc. Imeko Symposium of Moisture Measurement, Hungary, 1971, pp. 377-404.

131. L. V. L. Shastry, M. Srinivasan, and V. Subrahmanyan, J. Sci. Ind. Res. (India), 14A, 585 (1955).

132. C. F. M. Fryd and P. R. Koff, Analyst, 76, 25 (1951).

Chapter XII

MOISTURE IN SOILS, SANDS,
CONCRETE, AND SILICA AND SILICATES

1 MOISTURE IN SOILS

1.1 Introduction

In agriculture, horticulture, hydrology, civil engineering, etc., the moisture content of soils and clay minerals plays a very important role. The physicochemical properties of the various structures and materials utilized in these branches of science are in a large measure dependent on the moisture content as well as the distribution of moisture in them. It is, therefore, natural that a large number of methods and techniques have been developed and applied for estimating the moisture content of soils, sands, and clay minerals. In this section a short description of soil moisture and the correlation between the moisture content and the soil and suction characteristics have been briefly described, after which the various methods of sample preparation and techniques of measurement are discussed.

1.2 Definition of Soil Terms and Classification of Soil Moisture

Water is present in soils in many different forms, and it is extremely difficult to establish accurate boundaries between these different varieties of soil moisture. Moisture within an earth mass is neither uniformly distributed nor in a static equilibrium. This is especially true the closer a given point of an earth mass is to the surface, because of the influence of atmospheric factors such as changes in temperature and air pressure. At greater depths, soil moisture is more uniformly distributed through the earth mass and is more stable than when close to the surface of a mass. Hygroscopic moisture is attracted from the air by soil particles. The greater the internal surface of a soil the greater the quantity of moisture that may be attracted from the air by that soil. The soil moisture can be broadly classified as shown in Fig. 12.1.

A number of investigators have attempted to classify different conditions of water encountered in soils. According to Bouyoucos [1] and Puri [2] the three main divisions of soil moisture which have survived until now in soil literature are gravity water, capillary water, and hygroscopic water.

Gravity water is that water which exists when the soil is flooded and which moves downward when drainage is provided: Of the three forms of water it is the least important. Since the soil can be flooded with any amount of water, it has no fundamental relation to soil structure or other physical contents. We can, therefore, dismiss it as of only historical interest.

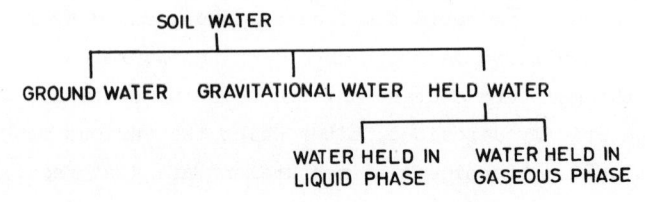

```
                        SOIL WATER
            ┌───────────────┼───────────────┐
      GROUND WATER   GRAVITATIONAL WATER   HELD WATER
                                        ┌──────┴──────┐
                                   WATER HELD IN   WATER HELD IN
                                   LIQUID PHASE    GASEOUS PHASE
```

FIG. 12.1. Broad classification of soil water.

1 MOISTURE IN SOILS

Capillary water exists in the pore spaces between the particles, filling them completely or partially.

Hygroscopic water is the water absorbed by the soil from a humid atmosphere. Thus all air-dried soils, not dried artificially by heating, contain hygroscopic water.

However, recently there has been a tendency to abandon the classification of soil moisture. It is tacitly assumed that all water is held by the same capillary force, the magnitude of which depends on the size of the pores. The hygroscopic and capillary moisture belong to different portions of the range; the former pertains to microcapillaries while the latter pertains to macrocapillaries. Yet the same surface tension effects operate in both cases. The soil-moisture relationship is expressed in a number of terms, some of which are described below.

1.2.1 Normal Moisture Capacity. It is defined as the minimum amount of water that is retained by absorption and film forces when water is free to move downward through a mass of uniform soil. To secure a numerical value for the normal moisture capacity, the average moisture content of the nearly uniform wetted portion of the soil column is taken. In obtaining this average, the moisture value for the first 1 or 2 in. of the soil is rejected. If a single value is to be taken, the moisture content of the middle portion is quite enough for the purpose. The normal moisture capacity would depend on the method of packing and sampling, the size and length of the soil columns, and the time allowed for moisture distribution. It is, therefore, not a fundamental constant for the soil, though it may give a rough indication of the moisture profile in a soil after rain or irrigation.

1.2.2 Field Capacity. Closely allied to normal moisture capacity is the field capacity, the moisture content of a soil determined usually from 1 to 5 days after a rain or on application of irrigation water. In subhumid regions where there is usually a dry layer below the wetted portion of the soil, the moisture conditions are comparable to those required for determining normal moisture capacity. In humid regions naturally the conditions are different.

1.2.3 Wilting Coefficient. The wilting coefficient may be defined as the limiting moisture content below which the plants are unable to draw water for their growth requirements. This may be determined directly by growing plants in a sealed container and withholding moisture content of the soil until they show signs of wilting. The moisture content of the soil at that point is determined in the usual way.

1.2.4. Moisture Equivalent. It is defined as the percentage of water retained by the soil material under specified conditions in a centrifugal field 1000 times the gravitational field of the earth. This moisture content is perhaps the best known and most widely used single-value soil property.

1.2.5 Plastic Limit. The plastic limit is defined as the lowest moisture content at which the soil can be rolled into a thread 1/8 in. in diameter without the thread breaking into pieces.

1.2.6 Liquid Limit. The liquid limit is defined as the moisture content at which 10 light shocks produced by striking the dish containing the wet sample against the hand will just close a groove made in the soil.

1.3 Movement of Moisture in the Liquid Phase

Although the water held in soil does not move freely under the action of gravity, it cannot be regarded as static. In general the movements which occur are slow, but considerable quantities may be transferred over long periods both in the liquid and vapor phases. The mechanism by which held water is retained in the soil and the

1 MOISTURE IN SOILS

factors which govern its movement are considered below. Held moisture moves in the liquid phase from regions of low suction to regions of higher suction. Vertical movements, however, are affected by gravity in the manner discussed below in connection with the distribution of moisture above the water table. An upward migration of moisture cannot occur in a soil between two regions having a vertical separation of a certain distance, say a few feet, unless the soil suction at the higher region exceeds that at the lower by the same distance. Neglecting this effect of gravity in a soil of uniform type, any local change in moisture content by destroying the suction equilibrium sets up movements of moisture tending to restore the moisture content at a uniform higher or lower value depending on the nature of the local changes. The movement and distribution of water in soils has been studied thoroughly by Croney et al. [3-4].

1.4 Moisture Movement in the Vapor Phase

If moist soil is placed in an enclosure, an equilibrium pressure depending on the temperature is again built up. The pressure is, however, less than the saturated pressure of water vapor at the same temperature as a result of the suction with which the soil water is held. The vapor pressure of soil water can, therefore, be expressed directly as a relative humidity. Vapor pressure of the moist soil increases with moisture content, but only at a low moisture content, i.e., at high suction, does the vapor pressure differ considerably from that of free water at the same temperature. It is well known that the vapor pressure of the moist clay minerals is lower than that of moist sand at the same moisture content and temperature.

At the moisture-content levels well below saturation, the air spaces inside the soil can provide continuous passage through which water may move in the form of vapor, and consequently it is possible for changes in the moisture content of soil to occur owing to the movement of water vapor from one region of the soil to another through these air channels. This movement is due to differences in the relative humidity of the water vapor in different parts of the soil.

Movement of water in the layers of soil above the capillary head is not fully understood. Water moves as liquid from water table up to the capillary head. According to the generally accepted view, the movement of moisture above the capillary head is in the form of vapor, and according to this view the only mode of transfer in the very unsaturated soil is by vapor diffusion. But volume diffusion is not able to account for the total observed transfer through the layers above the capillary head. Bodman and Colman [5] have confirmed that in certain soils capillary movement does take place at less than moisture-equivalent moisture content. Diffusion theory has been applied to the capillary movement of water in soil by Gardner and Widtsoe [6], Childs [7], and Ostashev [8]. In this theory it is assumed that the capillary flow of moisture is a phenomenon analogous to heat transmission in solids or to molecular diffusion in gases. Kirkham and Peng [9] have experimentally examined the diffusion theory and laws of capillary flow in soils. They have also discussed the fundamental assumptions involved in the diffusion theory.

In order to get more insight into the mechanism of water transfer, John [10] did a number of experiments to study the water movement in the top layers of soil. He concluded that wet front motion takes place only when ample molecules have been adsorbed by the soil surface. In other words, wet front motion takes place only when there is multimolecular adsorption on the surface of the soil particles above the wet front. If the humidity is low there will be no multimolecular adsorption, and hence no movement of wet front occurs.

1.5 Effect of Moisture on Soil Characteristics

The sorption of moisture by soils causes hydration, suction, and swelling. The individual particles of soils like silts and

1 MOISTURE IN SOILS

clays are bound together by films of water. The cohesive forces that arise from these films are thought to be of two types, viz., those due to surface tension forces at the air/water interfaces within the soil structure and those due to interaction between the soil particles or between the particles and water molecules. The forces causing this hydration, together with the surface tension at the air/water interfaces previously referred to, combine to produce a state of reduced pressure or suction in the water, the value of which is dependent on the moisture content of the soil. Another effect associated with particle hydration is the swelling of clay soils. At relatively short distances from the surfaces of clay particles the orienting and adsorbing forces acting on the water molecules are very strong, and the water is believed to be in the solid rather than in the liquid state (adsorbed water). As these adsorbed layers grow during the wetting of a clay, the effective solid volume associated with each particle increases, and if the layers are in contact with each other, the growth of the individual layers will be reflected in an increase in the total volume of the soil structure.

In practice the adsorbed water films in clay minerals grow in thickness until the suction pressure in the water becomes equal to the overburden pressure on the soil due either to self-loading or to externally applied loads. At the time when this equilibrium is reached, if the loading is increased, the adsorbed water films are reduced in thickness and settlement occurs. The loss of moisture due to evaporation or transpiration from vegetation takes place. Initially the volume of the soil decreases linearly with the amount of moisture lost; then the decrease in volume becomes less for a given loss of moisture and the relationship is nonlinear. The moisture content below which little shrinkage should take place is called the shrinkage limit. For most soils this value is fairly constant at about 12 to 14% equivalent to a voids content of about 30%. This voids content corresponds approximately to a theoretical porosity of 26%, which has been calculated to occur in a closely packed structure consisting of single-sized spherical particles. A constant shrinkage limit would, therefore, be expected for all soils if the shape of the particles were similar and independent of particle size. Another property acquired by soils as a result of water absorption is increase in plasticity, which in the soil is due to the lubricating effect of the water films between adjacent particles. Thus it is dependent on the factors which influence the area and thickness of these films, i.e., on the size and shape of the individual particles and the chemical nature of their surfaces. However, the thickness of the films is primarily dependent on the moisture content of the soil and the plasticity characteristics of soils are, therefore, generally studied by determining the amount of moisture required to bring them to arbitrarily defined states of plasticity.

1.6 Relationship between Moisture Content and Soil-Suction Characteristics

Soil moisture suction is due to the reduced pressure (below atmospheric pressure) imparted to hold water molecules by the forces retaining them. Soil suction increases continuously with decreasing moisture content over the entire moisture range. Its value rises from zero at saturation to many thousands of pounds per square inch in dry soil. This large variation makes the use of a logarithmic scale essential if the soil-suction/moisture-content relationship is being considered as a whole. Schofield [11] has endeavored to express the moisture content of soils in terms of free-energy relationships. If we suppose that water is held by soil against a suction force tending to displace it, the free energy can be expressed in terms of the height in centimeters of the equivalent water column.

The concept of suction provides a rational method for describing and controlling the moisture status of porous materials, particularly in the high humidity region. In many materials, moisture

1 MOISTURE IN SOILS

contents greatly increase with a small change in vapor pressure at relative humidities in excess of 95%; accurate measurement and control of vapor pressure directly is often so difficult as to preclude its use. On this scale, a column of water 10,000 cm high is equivalent to a relative vapor pressure of 0.993. The log of this expanded scale based on suction is Schofield's free energy F. Its ease of determination and control makes it most useful in many studies in which moisture in porous materials is involved. The moisture held at a fixed suction increases with clay mineral content. This is because the number of small channels in which water is retained and also the total surface area of the particles increases as the proportion of finer constituents increases. The structure and bulk density of the soil also appear to affect the soil-suction/moisture-content relationship. Since the surface tension at an air/water interface decreases slightly with increasing temperature, the suction of soil water decreases in a similar manner. The relationship between suction and moisture content at a depth of 6 in. is shown in Fig. 12.2.

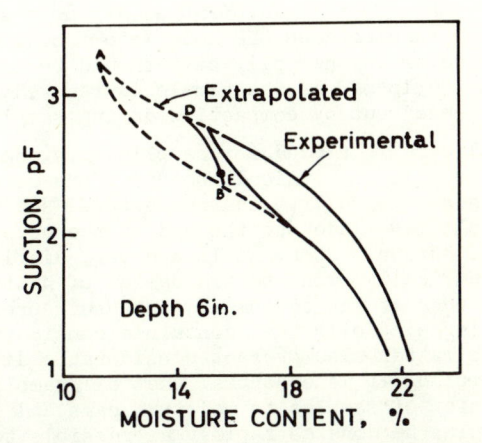

FIG. 12.2. Relationship between suction and moisture content.

XII. MOISTURE IN SOILS, SANDS, CONCRETE, ETC.

1.7 Sampling Procedures

1.7.1 Sample Collection. Proper sampling is the sine qua non for precise measurement of moisture content of soils. For determination of the moisture content of a soil, the soil sample taken should be representative of the soil. For the best samples, the soil should be homogenous and free from roots, organic matter, and stones. Any sampling scheme followed for the selection of a soil sample for the measurement of moisture content of the soil must also recognize changes in soil moisture with lateral and vertical displacement. The soil moisture may vary considerably from point to point, and the rate of change may be gradual in some cases and abrupt in others. Seldom are all these conditions met. Relatively little research has been reported on soil sampling, although sampling error may outweigh analytical errors. Because of the many complicated factors involved, it is impossible to give a set of inflexible sampling rules, but certain general guidelines to be observed are listed below [12-14]:

1. The technique and the equipment used for sample collection should be such that the samples do not lose or gain moisture or otherwise become altered or contaminated during sampling and transportation.

2. In sampling from a wet layer into a dry layer, care must be taken to keep the sampling equipment as dry as possible and to prevent water from running down the hole into the drier material. If there is free water in the soil, care should be taken to see that some water may not drip off as the sample is removed from the ground, or may not be squeezed out by compaction during sampling.

3. Special care is needed for sampling dry, hard, fine-textured sediments: it is difficult to drive the core barrels or to rotate the augers. When dry, coarse-textured sediments are sampled, the sample may slide out the end of the core barrel or auger as it is withdrawn. Stony soils are very difficult to sample, especially volumetrically owing to the danger of hitting a stone with the cutting edge of the equipment, because representative samples must be large. Soils that contain a considerable amount of roots and organic matter also present considerable difficulty. In soil moisture sampling it is essential that all sampling operations such as the transfer of samples to moisture cans and the weighing of the moist samples be done as rapidly as possible to prevent undue moisture losses.

1 MOISTURE IN SOILS

4. Special precautions should be taken with samples intended for the determination of the traces of water. Ordinary paper bags fitted with plastic (polyethylene) liners are suggested for field collection and transportation of samples to the laboratory. The amount of the sample should also be much more than the usual sample. The sample taken should not be allowed to come in contact with chemical fumes. Ammonia fumes are a common source of contamination.

1.7.2 Sampling Augers.

The ideal sampling tool is one that gives an uncontaminated, reproducible sampling unit of approximately uniform cross section to the desired depth. The simplest variety of auger in common use is the hand auger. Hand augers with shift extensions of aluminum pipe have been used in sampling to depths as great as 55 ft.

One of the most useful types of hand augers is the Orchard auger. It consists of a cylinder 3 in. in diameter and 9 in. long having a $4\frac{1}{2}$-ft extension pipe on the top and two curved cutting teeth on the bottom. Because the barrel is a solid cylinder, the sample is not as likely to become contaminated from the side of the test hole as with the Iowan or the posthole auger. For ease in sampling at depths greater than 5 ft, 3-ft extensions of $\frac{3}{4}$-in. aluminum pipe are added as needed.

To obtain a sample by the hand auger method, the auger is turned by its handle and forced downward into the material to be sampled. Usually about 3 in. of the material may be penetrated before the cylinder barrel is filled. The auger is then raised to the surface, and the sample is jarred loose from the auger barrel by hitting the barrel with a rubber-headed hammer. A good representative sample is thus obtained.

1.7.3 Sampling Tubes or Core Barrels.

Sampling tubes are more satisfactory, particularly for sampling surface horizons. A soil-sampling tube, core barrel, or drive sampler of some type offers an advantage in soil moisture sampling because volumetric samples can be obtained for calculating moisture content by volume. The King tube was developed in 1890, and Veihmeyer [15] developed modified soil tubes in 1929. Since that time many investigators have developed variations of this type of sampling equipment. Core samples provide uncontaminated samples if the equipment is kept clean. Oil should never be used on the samplers, and they should be kept free of dirt, rust, and moisture. A two-man crew is normally recommended for deep sampling. Depths up to 65 ft can be sampled. Sampling tubes are ineffective for sampling stony, dry, or sandy soils.

1.7.4 Porter Piston Sampler. The Porter piston sampler is one type of sampling equipment that has been very useful especially for sampling through loose or wet materials that tend to slough into the hole. This sampler is the retractable piston-drive sampler type. By means of a hand-operated 25-lb drop hammer a plugged sample section may be driven to the required depth by the addition of a 5-ft extension rod. The plug is then retracted, the sampler is driven a maximum of 2 ft, and the soil core is retained in brass insert liners contained in the 4-ft sample section. Further retraction of the plug helps retain the sample by forming a partial vacuum above it. After extraction, the liners with the soil cores are removed from the sample tube and capped and sealed for laboratory testing. The liners are brass cylinders, 1 in. in diameter and 6 in. in length. Two types of hardened-steel cutting points are available for use in different soils. This sampler is made from high-strength steel and is rugged and dependable. However, reasonable care must be taken in its use and storage to insure efficient operation and long life.

1.7.5 Pomona Core Barrel Sampler. Undisturbed soil samples can be collected by means of this barrel. The Pomona sampler (Fig. 12.3) consists of a core barrel 2 in. in inside diameter and 4 in. long, with extension tubes 1 in. in diameter and 5 ft long for sampling at depth. Brass cylinder liners, 2 in. in length, are used to retain the undisturbed core samples. The samples are removed from the core barrel by pushing on an attached plunger. A light drill rod or

1 MOISTURE IN SOILS 801

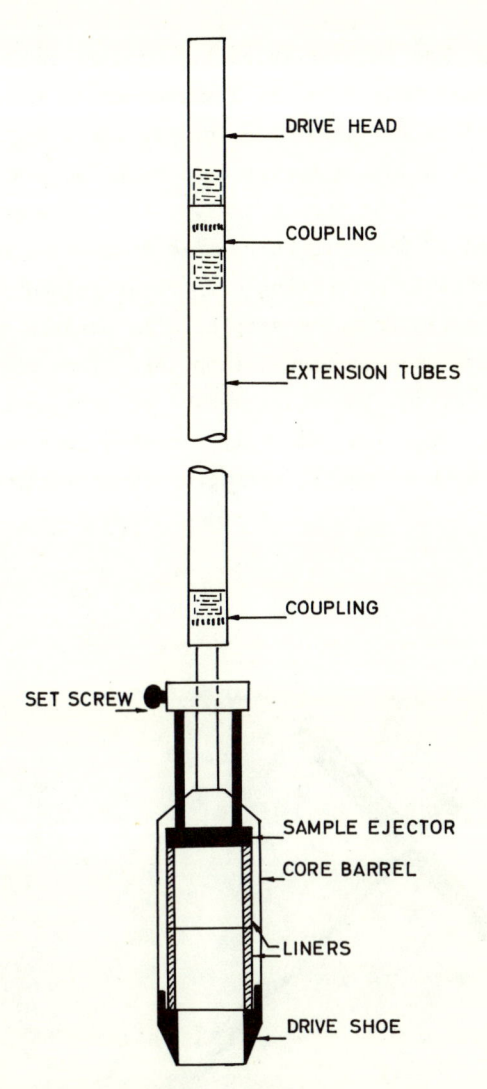

FIG. 12.3. Pomona open-drive sampler.

1-in. pipe may be used for extensions. The cores are collected by use of a 25-lb drop hammer.

1.7.6 Johnson Open-Drive Sampler. A simple and economical sampler for obtaining volumetric core samples from shallow depths was designed

802   XII. MOISTURE IN SOILS, SANDS, CONCRETE, ETC.

by Johnson [12]. The sampler is easily constructed and consists of
a thin-walled brass tube 2 in. in diameter and 6 in. long mounted
on the end of a 3-ft T handle of $\frac{3}{4}$-in. pipe (Fig. 12.4). Samples
are collected by a downward thrust on the handle and then are pushed
out of the core barrel by the central plunger. Because the inside
diameter and area of the core barrel are known, volumetric samples
may be easily obtained by cutting off predetermined lengths of the
core as it is removed from the sampler. In certain cases it is more
profitable to come back to manual sampling. Thus subsurface horizons
are best sampled by the laborious method of digging a pit, preparing
a fresh vertical face, and carefully removing sampling units at
suitable depths with a trowel, knife, or other instrument.

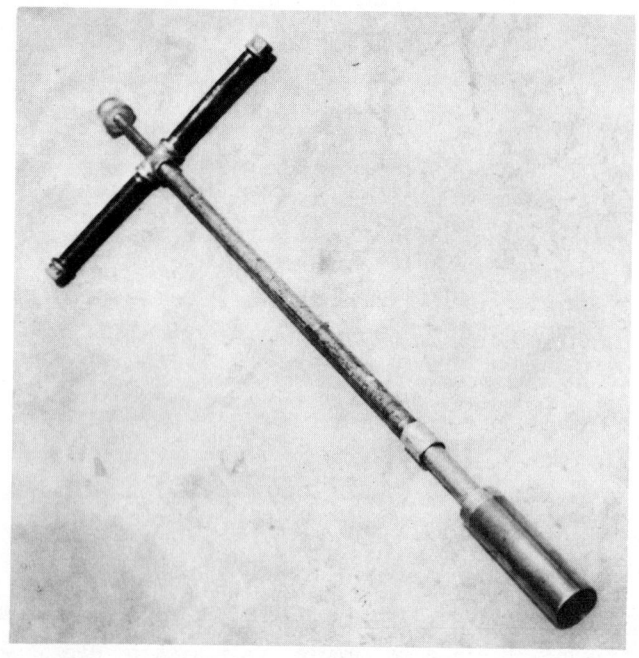

FIG. 12.4. Johnson open-drive sampler.

1 MOISTURE IN SOILS

1.7.7 Profile Sampling. If sampling in depth is desired at a specific point on the land surface, a pit should be dug to display the various soil layers (horizons) in vertical cross section (profile). Care should be exercised to avoid incorporation of soil from different horizons in the same sample. Careless sampling of lower horizons sometimes results in contamination with loose soil from upper layers. Failure to allow for transition zones between horizons is another source of error. The process is greatly simplified when a sample of the surface soil only is desired, as for instance, the ploughed layer of a cultivated field. In this case, any surface litter of undecomposed vegetative material is removed, and a sample of uniform cross section is taken to the desired depth. Such a single core or slice is defined as a sampling unit.

1.7.8 Area Sampling. Attempts to obtain a sample representative of a soil area must recognize differences in soil properties from point to point on the land surface. Estimates of mean values are obtained from a sample composited from several sampling units taken from a sampling area. These sampling units should represent the same soil horizons and should be uniform in volume and in cross section. It may be necessary to subdivide the sampling area and to prepare separate composite samples from each of these subdivisions if the area is of great extent or if it is not uniform with respect to soil properties, plant growth, or land management and use. The number of sampling units taken to make up the sample should be governed by the variability of the characteristic under test and the degree of accuracy of the estimate desired; 20 or 30 sampling units for surface horizons and 10 sampling units for lower horizons have been suggested as minimum values for some purposes. It should be stressed that these minima are arbitrary and, in many cases, a considerably greater number of sampling units may be advisable. The sampling units should be taken from sites chosen at random within the sampling area. The soil sample should be placed in a noncontaminating container. Closely woven cloth sacks are convenient for this purpose.

1.8 Methods of Moisture Measurement

A large number of methods and techniques have been developed for the measurement of moisture content of soils and similar materials. These can be broadly classified into the following categories:

1. gravimetric method
2. electrical resistance method
3. capacitance or dielectric method
4. hygrometric method
5. microwave method
6. neutron-scattering method
7. tensiometric method
8. thermal conductivity method
9. hygrophotographic method

In the following sections, these methods will be described only briefly since their working principles have been discussed in Chaps. II to V. However, microwave and neutron-scattering methods have been specially developed for the measurement of moisture content of soils and building structures and are therefore discussed in greater detail.

1.8.1 Gravimetric Method.

The gravimetric method for measuring soil moisture is the oldest (other than the ancient method of feeling the soil) but most widely used method for obtaining data on soil moisture. Because it is the only direct way of measuring soil moisture, it is required for calibrating the equipment used in the other methods. The gravimetric method involves collecting a soil sample, weighing the sample before and after drying it, and calculating its moisture content. Russel [13] and Whitney [14] describe some of the first scientific investigations of soil moisture measurement using gravimetric methods. Many types of sampling equipments as well as special drying ovens and balances have been developed for use with the gravimetric method and are described in a number of publications [16-20].

1 MOISTURE IN SOILS

The conventional method is to dry the soil in an oven at 105°C till constant weight is obtained. Unless the sample is air dry and finely ground, it is desirable to use a subsample of at least 25 g. The moisture content is expressed as a percentage of the oven-dry weight. But it does not represent all the water contained in it. For example, the clay mineral kaolinite loses about 16% by weight on heating from 105 to 800°C and other clay minerals have a smaller but still considerable loss on ignition. This loss on heating above 105°C measures the amount of -OH, water present in the clay mineral lattice. Organic matter content too is a source of hydrogen. A loss-in-weight determination on ignition of the whole soil will include loss due to organic matter. The error in moisture measurement due to evaporation of organic matter has been discussed by a number of researchers [21-25]. Indian Standards Institution, with a view to establishing uniform procedures for determination of different characteristics of soil and also for facilitating comparative studies of the results, has drawn standard specifications [26] for determination of moisture content of soils.

The disadvantage of the gravimetric method is the time and effort required to obtain data. It is a time-consuming job to collect the samples, especially from depths greater than a few feet, and to oven dry and weigh the many samples required for most projects. For many problems the sampling procedure may alter the area of experiment. Under these conditions the sampling may have to be refilled and packed. Soils are normally variable within an experimental area, and since two samples cannot be collected from the same point, slight variations of moisture content may be obtained.

1.8.2 Electrical Resistance Method. Among the many types of instruments proposed and developed for rapid and reliable soil moisture measurement, the most promising are those which involve measurement of electrical resistance in a porous material that is in moisture equilibrium with the solid soil enclosing it. The principle of electrical measurement of soil moisture was first reported by Whitney [14]. However, many years passed before successful electrical units were developed by Bouyoucos and Mick [27], Colman [28], Youker and Dreibelbis [29], Korty and Kohnke [30], and Colman and Hendrix [31].

A number of instruments based on the principle of the resistance variation in accordance with the change in moisture content have been developed. The most important of these are the ones developed by Colman and Hendrix [31] and Bouyoucos and Mick [27] which will be described here. The instrument consists of two parts, the soil unit and the meter unit. The soil unit is buried at the point where moisture measurement is required to be done, the electrical resistance is measured by the meter unit, and also the temperature. The Colman and Hendrix [31] moisture meter consists of a moisture-sensitive element and a thermally sensitive element enclosed in a monel case. The moisture-sensitive element is a sandwich composed of two monel screen electrodes separated by two thicknesses of the same cloth and wrapped around with three thicknesses of the same material (Fig. 12.5). The monel case of the soil unit consists of two identical half shells of 25-gauge metal which are die-formed to ensure dimensional accuracy and rigidness. When assembled and spotwelded along the flange edges, the case serves to compress the fiber glass uniformly, and good capillary contact between soil and fiber glass is ensured by the thinness of the metal.

The thermally sensitive element used by Colman and Hendrix [31] was a thermistor T which was found to be far superior to the wire-wound-type resistance thermometer. The resistance of the thermistor at 77°F was found to be about 1000 ohms. The high temperature coefficient of the thermistor used makes corrections for (Gd-M) lead-wire resistance unnecessary and improves the ease with which accurate temperature determinations of the soil block can be made.

The meter unit employed by Colman and Hendrix [31] was a battery-operated alternating-current ohmmeter (Fig. 12.6), which is entirely self-contained within a case approximately 8 in. wide, 12 in. long,

1 MOISTURE IN SOILS

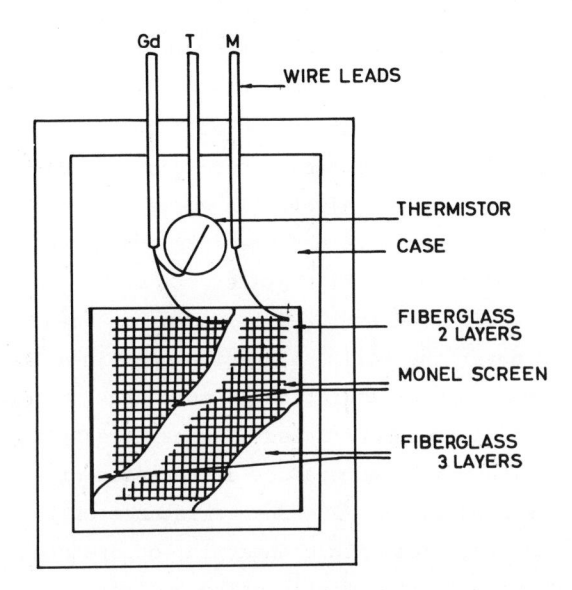

FIG. 12.5. Fiber glass block of the soil moisture meter.

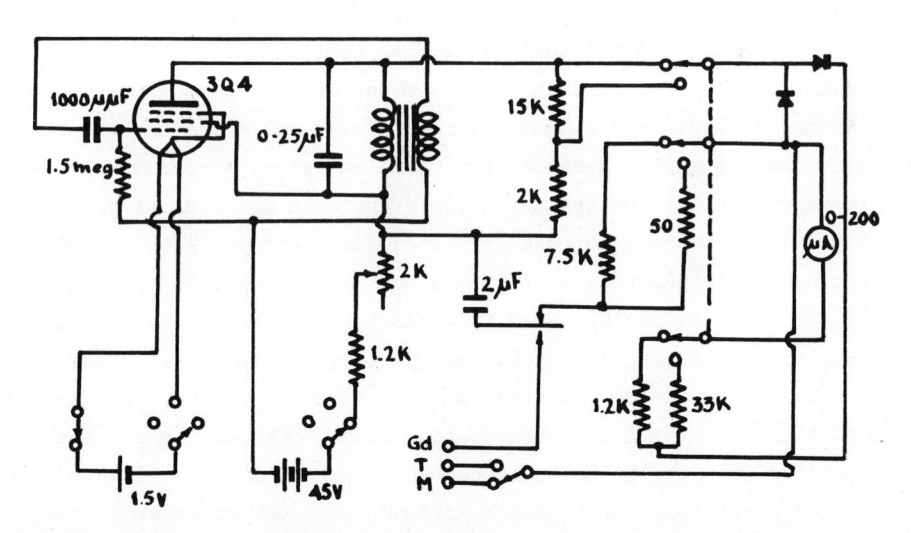

FIG. 12.6. Circuit diagram of Colman and Hendrix electrical resistance (fiber glass) moisture meter.

and 5 in. high. Alternating current of 90 Hz is generated by the vacuum tube oscillator and is fed into either the moisture- or the temperature-sensitive element of the soil unit. The emerging current is rectified so that its magnitude can be indicated directly on the microammeter. The calibration curve of the meter is such that at low scale, sensitive measurements can be made in the range from 50 to 10,000 ohms; at high scale, the useful range is from 8,000 to about 3,000,000 ohms. This range has been found to be sufficient to cover the soil unit resistances in all soils so far studied from pore-space saturation to moistures well below the wilting point. The meter is sufficiently sensitive to permit thermistor measurements with which temperature differences of about 1°F can be detected between 10 and 110°F. The resistance of the moisture-sensitive element varies with changes in temperature as well as in moisture content. For the most precise indication of moisture, therefore, the resistance measured must be converted to the resistance the element would have shown, under the same moisture conditions, at some standard temperature. In the soil studies made to date the standard temperature has been fixed at 60°F.

The Bouyoucos moisture meter also consists of two constituent units, i.e., the soil unit and the meter unit. The soil unit is a plaster of Paris block 1½ in. long, 1 in. wide, and 1 in. thick. Inside the blocks are imbedded two stainless steel screen electrodes running parallel to the long surface and held ¼ in. apart by a special technique. These electrodes are 15/16 in. long and 4/16 in. wide. The heart of this soil moisture-measuring apparatus is the plaster of Paris block itself. When this block is buried in the soil and left there, it practically becomes part of the soil. It absorbs moisture from the soil and gives it up to the soil very readily and quickly so that its moisture content tends to remain in constant equilibrium with the moisture content of the soil. The latest improved plaster of Paris blocks are made with 80% of water in the mixture (32 cm^3 of water to 40 g of pure plaster of Paris). This ratio of water to plaster of Paris increases the sensitivity of the blocks by increasing the pore space. When water in the pores is replaced by air, the electrical resistance is increased tremendously in the blocks. To prolong the life of the plaster of Paris blocks in the soil, they are treated with nylon resin. This treatment has great beneficial effects on the life of the blocks. In well-drained soils, treated blocks may last 10 years.

The electrical unit is either a special Wheatstone bridge or an alternating-current impedance meter, which is a modification of the Wheatstone bridge. It was built purposely for the plaster of Paris method. It has a resistance range from 0 to 50,000,000 ohms. It is powered by self-contained dry batteries which activate a 2000-Hz oscillator. It utilizes headphones, emits a powerful signal, has a large condenser to counteract capacitance effects, and is very sensitive and accurate. Although it is intended for research work, it can also be used for practical routine purposes. With this bridge even wire leads more than 200 ft long give accurate results. The alternating-current impedance meter is based on the principle of impedance matching as in the Wheatstone bridge. The electrical impedance offered by the plaster of Paris block is measured by this impedance meter (Fig. 12.7) which is calibrated to read both in percentage of available water and corresponding electrical resistance. It does not measure total moisture but only that portion of the moisture which is available to plants and which is in the range between field capacity and wilting point. The meter is hermetically sealed so as to be moisture-, humidity-, and dustproof. It uses mercury cells and transistors for long life. These are also kept in sealed compartments.

The relationship between the electrical resistance of the plaster of Paris block and the percentage of total moisture in the soil is shown in Fig. 12.8. Similarly, the calibration of the plaster of Paris block in terms of available soil moisture and directions for irrigating are shown in Fig. 12.9. It will be observed from Fig. 12.8 that the permanent wilting point which should be considered a narrow range rather than a specific value falls on the curves at about

FIG. 12.7. Plaster of Paris block and moisture meter.

100,000 ohms resistance. The field capacity falls on the curves at about 500 ohms resistance. At field capacity, the water is held by the soil with a tension equivalent to 0.3 atm pressure. In the range between field capacity and the wilting point of the plants is contained the total amount of water that is available to the plants. It is the function of the plaster of Paris block to measure this available water and to indicate when to irrigate. The different textured soils vary in their amount of available water reserve and in their water release, and therefore they have to be irrigated at different percentages of available water (Fig. 12.9).

Besides the fiber glass block and the plaster of Paris block described above a nylon soil-sensing unit has also been developed by Bouyoucos [1]. The special Wheatstone bridge used for the plaster

1 MOISTURE IN SOILS 811

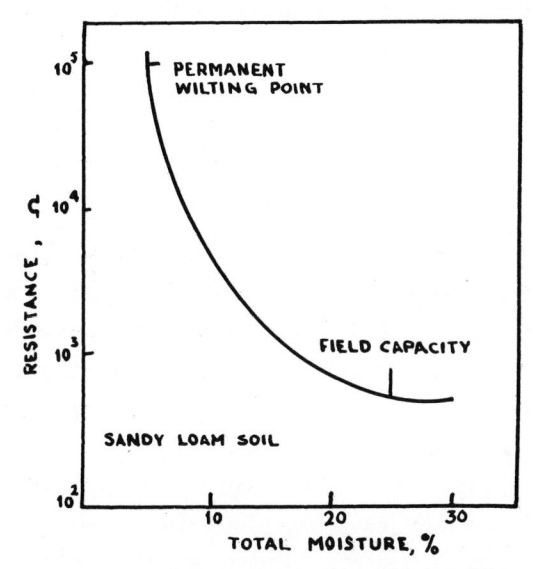

FIG. 12.8. Relationship between electrical resistance of plaster of Paris block and percentage of total moisture in sandy loam soil.

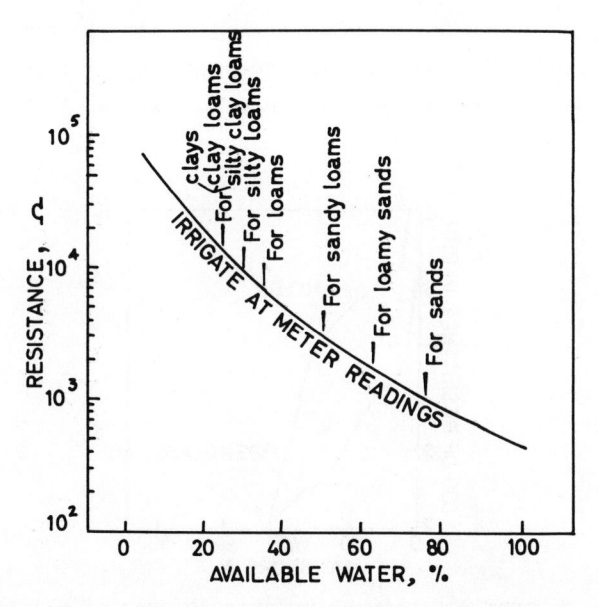

FIG. 12.9. The calibration of the plaster of Paris block in terms of available soil moisture and directions for irrigating.

812   XII. MOISTURE IN SOILS, SANDS, CONCRETE, ETC.

of Paris block method of determining soil moisture is also applicable to the nylon units. Any electrical or electronic arrangement which can measure the resistance accurately can be adopted for this purpose. With the exception of sand the nylon unit is capable of measuring moisture in all the different classes and types of soil varying greatly in salt and organic matter contents.

The nylon unit gives superior performance to the fiber glass unit. An advantageous feature of the nylon unit is its great sensitivity to changes of moisture at high levels of water content. For example, change of 23% moisture content causes the electrical resistance of the fiber glass unit to rise to only 1000 ohms, whereas the same increase in moisture content raises the electrical resistance of the nylon unit to more than 50,000 ohms (Fig. 12.10). In fact, the unit is more responsive and more sensitive to nearly saturated conditions than others thus developed so far. Because of this characteristic, the nylon unit is ideal for measuring the higher levels and greater ranges of soil moisture content.

The nylon unit has virtually no lag in its response to changes of soil moisture. Two factors are responsible for this: the extreme thinness of the unit and the extremely low water-holding power of the

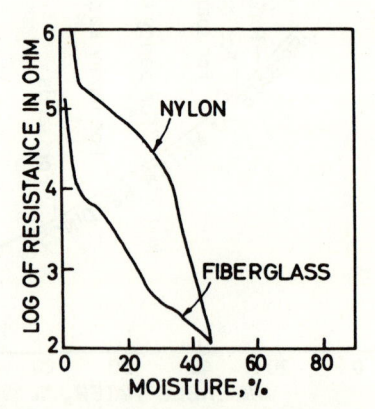

FIG. 12.10. Comparison between moisture curves of Clinton silt loam obtained by using nylon and fiber glass units.

nylon fabric. In addition, the nylon unit gives a more smoothly regular type of moisture curve than does the fiber glass unit. Admittedly, fiber glass has more buffering action than nylon, but this buffering action is not sufficient to overcome or eliminate the influence of salts; therefore, fiber glass has practically no advantage over nylon in this respect. Nylon is a very durable fabric and even when buried in the soil has a long life. In addition, nylon will not react chemically with soils and does not absorb and accumulate salts to any marked degree. Fiber glass, on the other hand, seems to react chemically with some soils and to absorb appreciable quantities of salts with consequent changes in the calibration of the unit.

The resistance read on the meter is converted to moisture-content values by means of a calibration chart. The calibration chart is prepared by correlation either in the field or in the laboratory from gravimetric moisture-content values and resistance readings for the soil in which the blocks are buried. Laboratory calibration consists of drying and intermittently weighing soil cores in which blocks have been inserted. Field calibration consists of taking gravimetric samples as close as possible to the blocks that have been buried in the field, and relating the moisture content of the sample to the measured resistance. For research investigations it is considered advisable to calibrate the soil unit for each soil installation [33].

The resistance of the moisture-sensitive element varies with changes in temperature as well as in moisture content. For the most precise indication of moisture, therefore, the resistance measured must be converted to the resistance the element would have shown under the same moisture conditions at some standard temperature. In the soil studies made to date the standard temperature is kept at 60°F. Soil temperatures can be measured at various depths using thermistors. Type F thermistors in which the resistance decreases from about 4000 to 2000 ohms for a rise of temperature from 0 to 20°C are used [3]. The element of each thermistor is usually encased in

814   XII. MOISTURE IN SOILS, SANDS, CONCRETE, ETC.

a protective brass sheath before being installed at the required depth. Resistances are measured by an alternating-current bridge. The accuracy of the thermistors used in this manner is found to be about ±0.25°C. Although not as accurate as thermocouples, thermistors have the advantage of being simpler to use.

The electrical resistance of the thermistor is related to temperature by direct calibration. For most accurate determination of soil moisture the nylon units as well as the fiber glass units should be calibrated for each soil and temperature correction applied to each measurement. The temperature response of a moisture-sensitive nylon unit is shown [32] in Fig. 12.11, which presents a standard temperature-correction chart.

For the most accurate determination of soil moisture, especially when nylon units are used, it is essential that the following precautions are observed.

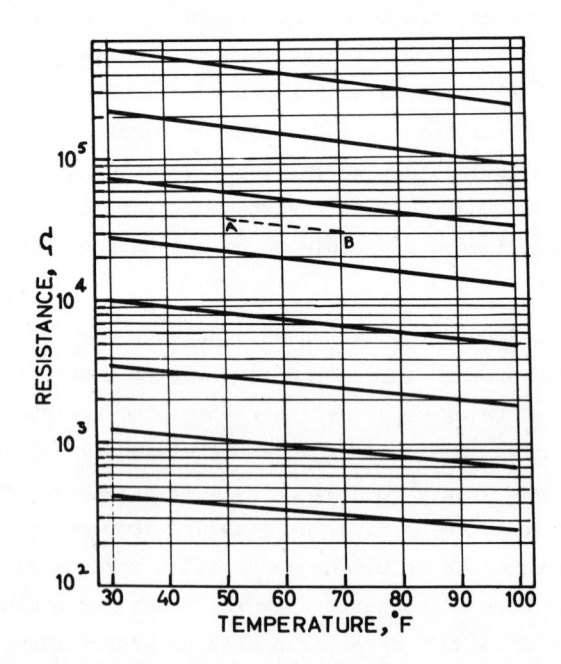

FIG. 12.11. Temperature-correction chart for nylon units.

1 MOISTURE IN SOILS

1. The units must be calibrated for each soil, and temperature corrections should be applied for each measurement.

2. An excess of water must be added to the soil when the units are calibrated. Not only must the sample be saturated, but there must be a thin film of excess water on top. This precaution ensures intimate contact between soil and absorbent surfaces.

3. Air trapped underneath the unit must be expelled by gentle pressure on the unit and by tapping the pan.

4. After the unit has been properly settled in the pan and water added, the pan is covered and allowed to stand for from 5 to 10 hr to allow establishment of chemical and physical equilibriums.

5. For calibration purposes, the unit should be equipped with flexible lead wires so that handling after equilibrium is established and will not disturb the absorption unit or the sample. Clamping the leads to the pan ensures that the unit will not be disturbed during subsequent manipulation.

6. It is recommended that calibration measurements be made on the second, rather than on the first, drying cycle. After every wetting, however, the air must be expelled from under the unit and the pan contents settled by gentle tapping.

7. Drying should not be hastened, but should be allowed to take place slowly.

1.8.3 Capacitance or Dielectric Method. The procedure for obtaining the electrical capacity is similar to that of determining the electrical resistance of the plaster of Paris blocks. The arrangement with electrodes serving as electrical condenser plates is shown [34] in Fig. 12.12, where P_1 and P_2 are two bare condenser plates exposed to the soil, and p_1 and p_2 are the electrical connections to the probe. For the measurement of capacitance a Wheatstone bridge is used which is balanced with respect to both capacity and resistance. The electrical capacity C_4 of the two-electrode plaster of Paris block from the settings of the Wheatstone bridge is given by the following equation:

$$C_4 = \frac{R_1}{R_2} C_3 \qquad (12.1)$$

For determining the value of the variable condenser C_3 in one of the arms of the Wheatstone bridge (Fig. 12.13) a four-decade capacitor

816 XII. MOISTURE IN SOILS, SANDS, CONCRETE, ETC.

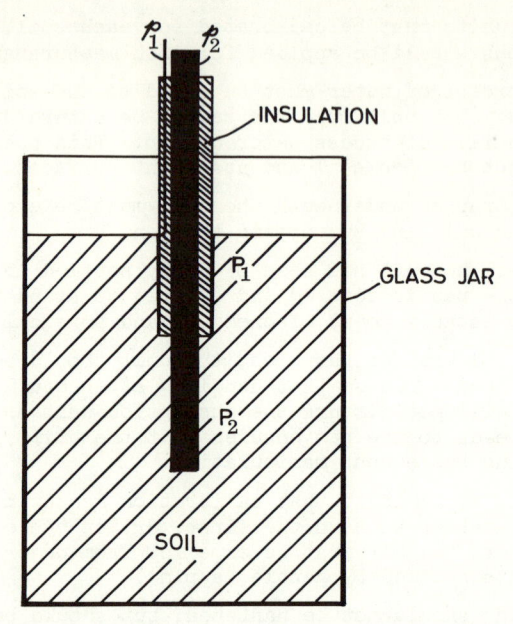

FIG. 12.12. Capacitance probe for soil moisture meter.

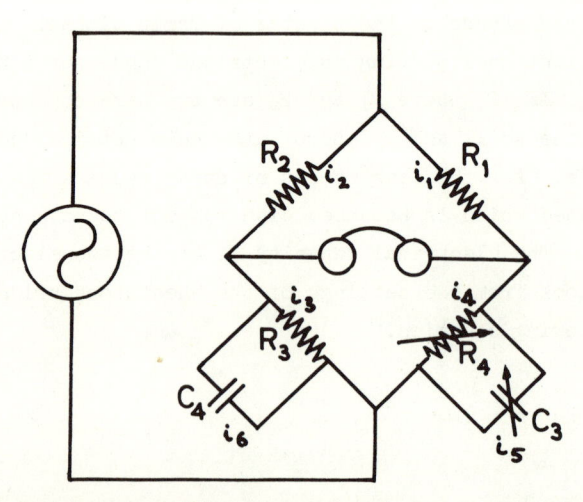

FIG. 12.13. Wheatstone bridge for determining the electrical capacity of the plaster of Paris block.

1 MOISTURE IN SOILS

is used having a range of electrical capacity extending from 0.0001 to 1.0 µF in steps of 0.0001 µF. It will thus be observed that the value of the capacitance C_3 can be determined very accurately.

The capacitance is determined at a frequency of 1000 Hz with the ac Wheatstone bridge. The probe consists of a conducting metal cylinder embedded in a round insulating rod, through the center of which passes a metal rod making up the other plate of the condenser. No attempt is made to insulate either of the electrodes to prevent the flow of current between them in the soil. When making the capacitance determinations at high frequencies and where low cost and small bulk of the apparatus are important considerations, one must make the tuning sharp enough to enable a reliable capacitance determination. No such difficulties arise, however, if lower frequencies are employed. Omission of the insulation not only renders the probe much more sturdy and simple, but makes the electrical capacitance, as determined by this method, much more sensitive to changes of soil moisture content. The cause of the increased sensitivity is the polarization that occurs in the soil at the surface of the exposed condenser plates. The greatly increased sensitivity of the condenser makes the determination relatively independent of the length of the leads. Of course, the exposure of the electrodes directly to the soil would be expected to make the determination of the soil solution; but this difficulty should be small for most agricultural soils. A detailed account of the experimental procedure using the electrical capacity of the plaster of Paris block is given by Anderson and co-workers [35-37].

On theoretical grounds Childs [38] concluded that the electrical capacitance of a plaster of Paris block will not serve as a reliable indicator of soil moisture content. According to him the apparent large increase in capacitance is due to the reduction in the resistance of the material of the plaster of Paris block, irrespective of any increase in dielectric constant. This is supported by the fact that the range of variation of capacitance of the plaster of Paris expressed as the ratio of the maximum to the minimum capacitance obtainable with change of moisture content exceeds the dielectric constant of water.

Although some of the arguments put forward by Childs [38] are essentially correct, others are based on a number of experimentally unfounded assumptions. Thus the relative magnitudes of the various resistances and capacitances involved are not correct, as is the effect of the assumed presence of poor electrode contact when the dielectric is leaky. Childs makes the assumption that the sheath resistance (by which he means the contact resistance) between the electrode and the soil is infinite. This is not supported by experiment. If it were, the dc resistance measured across the two electrodes would be infinite. He also assumes that the constant conditions between the electrodes and the porous material of the blocks are poor and a source of trouble. Anderson and Edlefsen [37] find this contrary to experimental observations. Even if the contact were poor, it would make no difference in using the blocks as the indicators of soil moisture content so long as the contact is reproducible with change of moisture content. It has been shown [35] that the reproducibility of the contact is excellent when the contact resistance is eliminated by the use of four electrodes. It is possible that the direct contact between the electrodes and the soil may not be quite so good as the contact between the electrodes and the porous block material, but nevertheless there is a good reproducibility as shown by the data reported by the investigators referred to above. It seems likely that this reproducibility can be improved with further development in the technique.

Figure 12.14 shows how the electrical capacitance of the soil depends on the moisture content (based on grams of moisture per gram of dry soil) of the soil. The moisture equivalent and the permanent wilting percentage of the soil are indicated on the curve. Anderson [34] has determined the relationship between electrical capacity and moisture contents of a series of soil samples extending in their range of moisture contents from a moisture equivalent of 6.7 to 29.1%

1 MOISTURE IN SOILS

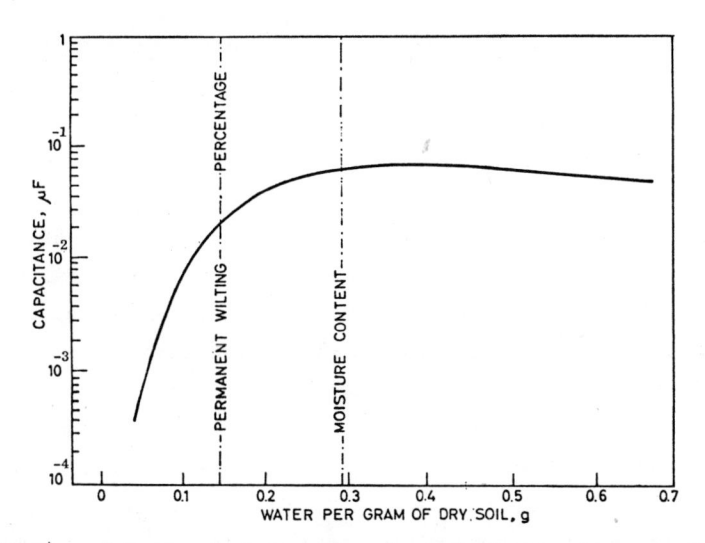

FIG. 12.14. Relation between the electrical capacitance of moist soil and moisture content expressed as moisture per gram of dry soil.

and has found a similar relationship. The samples of the soil experimented upon were obtained from locations more than 15 miles apart. He further showed that if, instead of plotting the capacitance as a function of the moisture content expressed as grams of moisture per unit weight of the dry soil, a graph is plotted between the capacitance and the moisture content expressed as grams of moisture per unit volume of the soil, similar curves are obtained. All the data so far available seem to indicate that there is a universality or identity in the shape of the curve showing the relation between the electrical capacitance of the soil and the soil moisture content over the range of moisture readily available to plants when the permanent wilting percentage is taken as the origin.

The curve of Fig. 12.14 discloses a much greater change of capacitance in going from a dry soil to a wet soil than what might be expected from the fact that the dielectric constant of dry soil is about 5.0, whereas that of water is about 80. On this basis the condenser when placed in pure water should possess only about 16 times the capacitance it possesses when placed in the dry soil. It is apparent that there is an enormous change in capacitance in going from relatively dry soil to wet soil. The anomalous increase in capacitance can be explained on the basis of interfacial polarization on the surface of the soil particles, in addition to the electrode polarization caused by the fact that the electrodes are not insulated. Evidence indicates [39-45] that the capacitance measured across the terminals of a condenser, the plates of which are uninsulated, and which is immersed in a suspension of colloidal or solid particles, may be considered as the sum of two components: (1) the capacitance of the material itself between the plates and (2) the component due to electrode polarization. The first component increases with decrease in frequency and is usually many times higher at low frequencies than at higher ones. The reason possibly is an abnormal polarizability of the multimolecular film of oriented water molecules at the interfaces of the soil particles in suspension. This water can be more or less closely identified with adsorbed or bound water. In various mechanical characteristics this layer seems to behave more like a solid than like a fluid.

It seems reasonable, therefore, to assume that under the influence of an electric field the layer would show polarizability (properties) approximating those of many dipolar crystalline solids. That is, it is less polarizable than water at very high frequencies, but becomes more highly polarizable than water when the frequency is low. This capacitance should, therefore, also increase with the total interfacial surface on the soil particles. Likewise, the second component increases with decreasing frequency. In those cases where a low frequency of the order of 1000 Hz is used for exciting the Wheatstone bridge used for the measurement of the capacitance formed by the soil as a dielectric between the blocks of plaster of Paris, the changes in the measured capacitance are out of proportion with a corresponding change in the moisture content of the soil due to the polarization action which is quite prominent at the low frequency used.

For most of the soils it is indeed unnecessary to recalibrate the behavior of the capacitance soil moisture-content curves because a given block remains virtually the same in shape over the entire range of readily available moisture in plants, no matter in what soil the blocks are placed. Of course, for example, if the calibration curves obtained by embedding the block in a sandy soil of which the range of readily available water was small had to be used to predict soil moisture content when the block was embedded in a clay soil, the curve should be expanded like an accordion along the moisture-content axis so as to fill exactly the new and greater range of moisture readily available to plants found for the clay soil. In other words, if a calibration curve obtained from the soil of one texture is to be used for a similar soil of a different texture, then the curve must be expanded or contracted along the moisture-content axis by an amount equal to the ratio of the range of readily available moisture of the second soil to the range of readily available moisture of the first soil. One need not therefore calibrate the probe for all soil textures in order to use the probe as an indicator of soil moisture content.

Measurement of moisture content of soils, sands, etc., can also be made by placing the electrodes directly in the soil or in some other medium the moisture content of which is sought instead of placing them in porous materials, and then there is no longer any need of having the roots of actively transpiring plants permeate the soil up to the porous block or electrode boundary. Thus one can determine the electrical resistance or the electrical capacitance as indicators of soil moisture content immediately on inserting the electrodes in the soil. The application of fertilizer, bacterial action, and other organic processes tend to increase the water-soluble impurity concentration. In general, therefore, the use of the measurement of electrical conductivity as an indication of moisture content of a moisture-sorbing material is unreliable if the concentration of soluble impurities is subject to variations. The behavior of the resistance is too erratic in such conditions and has not been found to give accurate results [36]. On the other hand, the electrical capacitance between two electrodes immersed in the soil might be expected to show more promise as a soil moisture-content indicator. With suitable experimental techniques, a stable and reproducible behavior of this electrical characteristic does not seem to depend so greatly on maintaining a good electrical contact between the soil and the electrodes or on a variation in salt concentration in the soil solution, because the permittivity of water is not significantly affected by dissolved substances usually found in soils. The values of permittivity for water vary from 78 to 81 over the temperature range 15 to 25°C and are independent of frequency from 50 to 1000 Hz. This high value of permittivity is accounted for by the dipolar nature of the H_2O molecule, which creates a statistical degree of alignment of the molecules superposed on the normal thermal oscillations when water is subjected to an applied electrical field.

An electronic moisture meter based on the dielectric variation of moisture content of hygroscopic materials has been developed by Pande [39]. A special needle-type electrode system for the measurement of moisture content of soils, sands, and similar granular materials has also been designed and constructed. The needle-type electrode consists of two rows of needles (separated by a distance of 3 to 4 in.) insulated from each other, which are mounted on a hylam insulating material to form a capacitance, the rows behaving like parallel plates of a condenser. The sample of soil forms the dielectric between these needle electrodes. The entire setup for the measurement of moisture content of soil and like materials is shown in Fig. 12.15. The reading of the meter is calibrated to read moisture content values directly, no graph or charts are required. A good agreement between the results obtained by this technique and those obtained by the oven-drying method was found. This meter and electrode system can be conveniently used for irrigation purposes in agricultural fields, the more so the transistorized version [40] of this moisture meter which is portable and fit for field operations. The use of very high-frequency (10 to 13 MHz) crystal-controlled radiofrequency oscillation considerably reduces the anomalous effect

1 MOISTURE IN SOILS

FIG. 12.15. SRI moisture meter with needle-type electrode.

observed in the case of plaster of Paris blocks when very low frequencies are employed for capacitance measurement.

The capacitance method of measurement of soil moisture content is a very promising method whereby it would seem unnecessary to calibrate the condenser over the entire range of soil moisture content for every soil on which the method is used. With decrease of soil moisture content, the electrical capacity of the block begins to drop from a rather high value to a little above the permanent wilting percentage of the soil. The dependence of the electrical capacity of the blocks on the soil moisture content shows an excellent reproducibility and an absence of lag in response of the blocks to changes of soil moisture content. Results indicate that the electrical capacity of the plaster of Paris block over the entire range of moisture content readily available to plants can serve as a practical indicator of the soil moisture content in a body of soil where the blocks can be buried and where the changes in soil moisture content are caused by the removal of the moisture by the roots of actively transpiring plants. When the behavior of the dependence of the electrical capacity is compared with the dependence of the electrical resistance of these blocks on soil moisture content, it is observed that at about the moisture equivalent value of the soil the electrical capacity is high and the resistance is low. As the moisture content decreases, the capacity decreases and the resistance increases rapidly until very close to the permanent wilting percentage of the soil. Here the electrical capacity reaches a seemingly constant value, whereas the resistance continues to increase asymptotically to very high values. It is not claimed that the measured capacitances of the blocks are absolute measurements of capacitance. They are empirical, though as such are surprisingly reproducible when the experimental procedure described in the foregoing section is adhered to. That they are reproducible is all that is necessary when using the capacitance method as an indicator of soil moisture content. However, this method has certain drawbacks which are discussed below.

The main objection to the determination of moisture of powdered and granular materials by the capacitance method is the spread of values due to variations in particle size and packing. Several moisture meters designed and developed for specific materials and based on the measurement of capacitance with high-frequency current have been described in the literature [46-51]. These instruments have been found to be satisfactory only with relatively low moisture-content samples. In order to overcome the drawback of these moisture meters, especially as applied to soils and similar materials, an experimental and theoretical study of in situ measurement of moisture in soil and similar substances by the fringe capacitance method has been conducted by Thomas [52] covering a range of moisture contents from about 0 to 45%. It was found that from 0 to about 10% there is a linear relationship between moisture content and the fringe capacitance (or real dielectric constant) of the test material; however, between about 5 and 45%, the relationship was such that the moisture content was a linear function of the logarithm of the fringe capacitance (or real dielectric constant). Over most of the range the standard error was approximately equal to 0.5% moisture content (volume basis).

There are reasons for expecting that above a fairly low value of moisture content, the real relative permittivity of soil, measured

1   MOISTURE IN SOILS

at very high frequency (vhf) as a function of moisture content (volume basis), will be largely independent of the type of soil concerned. The experimental results presented by Thomas [52] covering a range of moisture content from 1 to 45% and a range of types of simulated and natural soils provide reasonably conclusive evidence that this expectation is fulfilled. This conclusion greatly enhances the value of the very high-frequency capacitance measurement method for the determination of moisture content.

Agricultural, horticultural, and soil scientists prefer volume-basis values as being more directly related to the water available to plants, and hence variable; most directly involved in the relationship of permittivity to moisture content is the volume fraction of water, not the proportion by weight. The pore space of practically all natural soils in the dry conduction falls within the range of 40 to 60%, an average soil having about 50% with permittivity of 2.6. Thus permittivity of dry natural soils may have a range of dispersion about the mean of about 10% due to this factor. This dependency becomes insignificant at higher moisture content values because of the increasingly predominating effect of the high value of permittivity of water.

The real permittivities of the solid organic and inorganic constituents of soil in the dry condition range from about 2.0 to 5.0, but dry soil with a pore space of about 50% usually has a permittivity which does not differ much from 2.6. It is, therefore, to be expected that quite a low concentration of water (having a dielectric constant of 81) in soil will cause an appreciable increase in the effective permittivity provided the freedom of the H_2O (water) molecules to turn in the direction of the applied field is not reduced by surface forces. The experimental procedure adopted by Thomas [52] consists of the following:

1. Design of a novel probe for in situ measurements
2. Adoption of Wayne-Kerr vhf admittance bridge using 30 MHz for capacitance measurement

3. Statistical analysis of the data obtained

Finally he has compared the experimental results obtained with theoretical calculations as well as with the results obtained by other investigators under similar circumstances.

The primary object of the method is to establish an in situ measurement of soil moisture. This involves the design of a test probe incorporating electrodes which can be inserted in otherwise undisturbed soil and enable measurements to be made of the local condition down to depths of 3 ft or more under favorable conditions. The probe has coplanar stainless (nonmag) steel electrodes fixed to the sides of the tip which is wedge shaped (Fig. 12.16). The fringe capacitance of this probe was 1.65 pF over the range 0 to 10% moisture content (volume basis). The in situ probe together with $\lambda/2$ length of twin-core vhf cable had an effective capacitance of 12.5 pF in air and about 19 pF in soil having 10% moisture content. The effective capacitance of the probe was, however, not critical. By doubling the electrode width dimension, the fringe capacitance could be doubled with probably only a small change in the total effective capacitance and this would improve the accuracy. Wooden handles were provided to the upper end of the probe so that it could be pushed directly into the softer soil. However, for harder soils, it was desirable first to make a hole with an auger or similar device. The electrodes were brought into good contact with the contiguous soil by maintaining a moderate pressure on the top of the probe. To make the capacitance test, the test probe can be pushed vertically downward into the material. The in situ field tests were made at depths of about 3 ft. The soil down to a depth of about 20 in. was first removed, and the probe was then pushed down vertically to a further depth of 6 in. (Fig. 12.17). Immediately after making the measurement and withdrawing the test probe three separate cylindrical cores of the material were removed from the neighborhood of the hole by means of a soil sampler. These test cores were then used for the gravimetric determination of wet density and moisture content.

1 MOISTURE IN SOILS

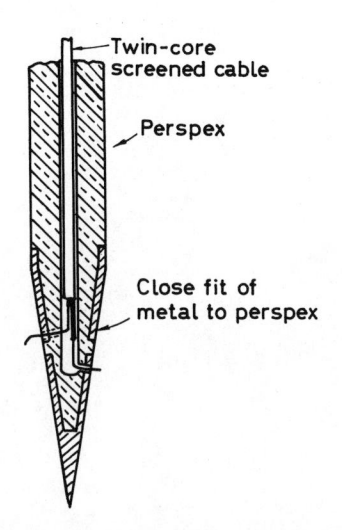

FIG. 12.16. Probe for soil moisture measurement by fringe capacitance method.

Connection to the measurement circuit was made by means of a 75-ohm, balanced twin-feeder vhf polythene-insulated and screened cable of special quality. The use of twin-core instead of the more commonly used coaxial cable is found to be beneficial in eliminating the effects of stray earth capacitance and so improves the accuracy and sensitivity of the measurement. The procedure followed is to obtain a zero reading with the probe in air and to measure the increase of capacitance when the probe is inserted into the soil or material under test.

The experimental results [52] (Fig. 12.18) provide clear evidence of the existence of the functional relationship between moisture percentage by volume and change in capacitance, which is independent of soil type within the limits of experimental error and sampling variability associated with the materials used in the study.

The fact that the variation of permittivity with moisture content is shown by experiment to be approximately independent of soil type is in agreement with theoretical expectation [53-54]. Exceptions to this are probably soils consisting mainly of colloidal clay

828 XII. MOISTURE IN SOILS, SANDS, CONCRETE, ETC.

FIG. 12.17. Photograph of the probe showing insertion into soil.

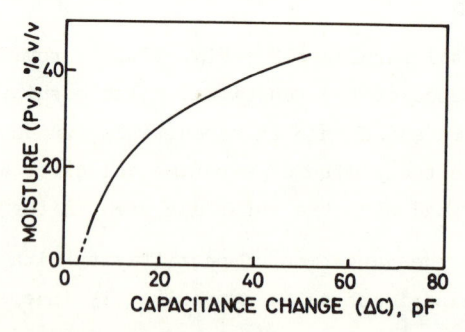

FIG. 12.18. Functional relationship between moisture content, volume basis, and change of capacitance according to the experimental measurements.

1 MOISTURE IN SOILS

(particle dimension, say, of 10^{-5} cm) which has exceptionally high internal surface area per unit volume, or organic soils which are known to have an exceptionally high heat of wetting, since these materials possess high physical and chemical sorption properties, respectively. The functional relationship obtained from theoretical considerations between moisture content by volume and log of capacitance change is shown in Fig. 12.19, from which it will be observed that a linear relationship exists.

From exhaustive theoretical and experimental studies it has been concluded by Thomas [52] that the portable industrial-type capacitance meter for use with the design of probe used in the investigation should be a dual-range instrument. The lower range should cover the measurement of a capacitance change from 0 to about 7 pF in the presence of shunt conductance of about 1 mmho (1000 ohms resistance). The upper range should cover measurement of capacitance changes of about 5 to 50 pF in the presence of shunt conductance of a maximum of about 20 mmho (50 ohms resistance). However, the commercial exploitation of the methods depends on the development of a relatively inexpensive vhf capacitance-measuring instrument. Besides the method developed by Thomas described here two electrodes for in situ measurements in soil had been earlier developed by Anderson [34] and de Plater [54]. These electrodes were parallel plates of stainless steel mounted on an insulating handle, and the electronic

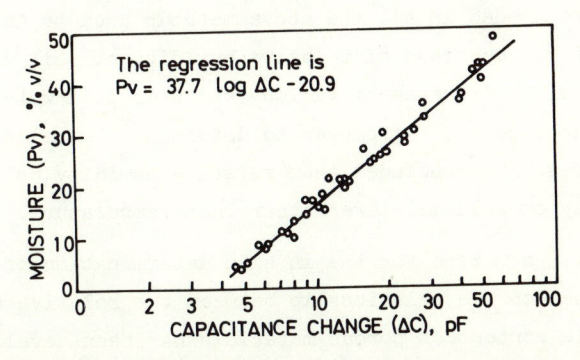

FIG. 12.19. Functional relationship between pV and C.

circuits used were Schering Bridge and transformer coupled bridge, both operating at 1000 Hz. These methods were found to be unsatisfactory due to the low frequency employed and uncertainty of contacts between the electrode and the soil.

1.8.4 Hygrometric Method. A large number of investigators have studied the relationship between relative humidity of air and moisture content of soil and similar hygroscopic materials. Labedeff [55] was one of the first to determine relative humidity of soil/air. He found it by means of the hair hygrometer. He concluded that (1) when soil moisture is less than its hygroscopy, then relative humidity is less than 100%; (2) the drier the soil the less the relative humidity; and (3) when soil moisture content is constant with temperature, relative humidity of soil/air increases. Puri [2] determined the relation between soil moisture and relative humidity of soil/air by keeping soil in a particular humidity in a closed container or by passing air of particular humidity through the soil for a long time. He came to the conclusion that the moisture and humidity relationships in a vast number of soils give almost identical isotherms irrespective of the nature and origin of the soil. Rocha [56] determined the relative humidity in a cavity by means of strain produced in wood prisms with the absorption of moisture from air. He stated that relative humidity will decrease with decrease of temperature if the medium is an adsorbing one. Accurate calibrations are needed in all the above methods used to determine relative humidity. Decrease of relative humidity of soil/air can also occur due to the presence of various salts in soil water. Fukuda [57] used an electric hygrometer to determine the relative humidity in soil pores. He concluded that relative humidity below 100% depends mainly on soil moisture rather than temperature.

Recently a method for the in situ determination of moisture content based on the relationship between the relative humidity and the moisture content of porous materials has been developed by De Costro [58]. He employed a vibrating-wire telehygrometer, whose working principle is based on the length variations effected in certain hygroscopic materials upon the absorbing or losing of moisture. He measured these strains by means of a vibrating wire whose vibration when excited by an electromagnet has a frequency which is a function of the stress in the wire. The results obtained by him can be summarized as follows:

1. The field of application of the method is limited to moisture contents corresponding to relative humidities below saturation.

2. Results yielded by vibrating-wire telehygrometers are only accurate below 80 to 90% relative humidity. Accuracy can be improved if it is known whether the phenomenon in question is drying or wetting. Calibration is more accurate for small humidity variations.

3. In spite of its stability in time and durability, the hygrometric material so far used in telehygrometers has the disadvantages of requiring an individual calibration for each instrument and of presenting a certain hysteresis and slow response.

A thermocouple psychrometer for measuring the relative vapor pressure of water in liquid or porous materials with a sensitivity of ±0.002% rh at high humidities has been developed by Richards [59]. This instrument measures the combined water-binding effect of the solid phase and dissolved solute in soil and was developed for measuring the availability of soil water for plant growth. For this purpose, high accuracy is required because the economic growth of crops is limited to the relative humidity range in soil above 99%. A number of researchers [58-61] have worked on the development of such an instrument, and since it is supposed to be one of the most accurate methods of measuring the relative humidity in soil in high humidity, a brief description of the working principle, constructional details, and some of the applications is given here.

Figure 12.20 shows the assembly of the thermocouple with soil sample installed in a thermostat. An insulating cover, a, helps to maintain the samples at a constant temperature in the liquid bath, b. The masonite cover, c, supports a number of thin-walled brass test tubes, d. The thermocouple mount guides the couple into the

832   XII. MOISTURE IN SOILS, SANDS, CONCRETE, ETC.

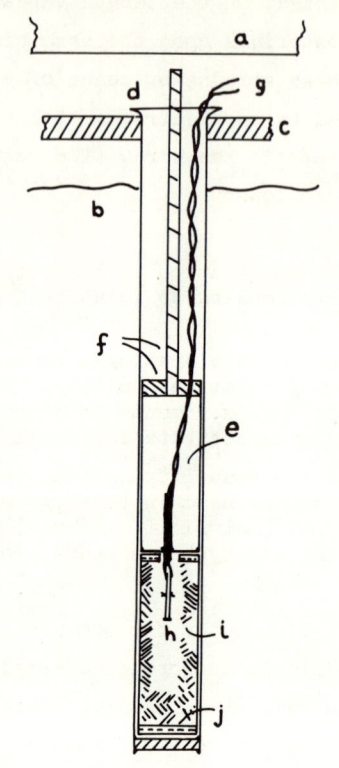

FIG. 12.20. Richards thermocouple psychrometer.

sample container and makes it convenient to shift the couple from one sample to another. The mount consists of a cylinder of thin-walled brass, e, closed at the lower end with a disk of copper and a copper tube assembled with soft solder. The handle, f, is made of rigid plastic. The lead wires, g, have seven strands of 36-gauge bare copper and have vinyl insulation of 0.1 cm outside diameter. Soft solder with low thermal electromotive force is used to join bare chromel and constantan wire, 25 μm in diameter and to attach the silver cylinder, h, which is of 0.185 cm outside diameter with a wall 0.018 cm thick and 0.051 cm high. The thermocouple resistance is 20 ohms. Soil samples are prepared by filling the sample container with a closely fitting soil core and closing the ends with solid caps.

1   MOISTURE IN SOILS   833

        The thermocouple output can be measured with a precision of
0.01 μV by a potentiometer arrangement like that described by Teele
and Schuhmann [60]. The thermocouple is calibrated by taking readings on standard osmotic solutions supported on filter paper. After
placing the sample in the bath, the thermocouple with its water
droplet is lowered into place. The vapor seal is made by a coating
of petroleum jelly on the end of the thermocouple mount. Air-pressure
adjustment in the sample chamber takes place in the clearance between
wires and tube of the mount. A steady electrical output is usually
obtained in 10 to 30 min if the vapor condition of the sample is
steady. A longer time is required at higher relative vapor pressure
[61].

1.8.5 Microwave Method. The theory and design features and constructional details of the microwave method have been described in
Chap. V and the literature [62-63]; therefore, only the application
of the moisture meter for the measurement of moisture content of
soils and other building material is described here. The suggested
method of using the microwave moisture meter to determine the moisture content of building material is as follows: The two transmitting
and receiving units are set up facing each other and spaced at a distance equal to the thickness of the material or wall to be tested.
The attenuator is adjusted and when center scale indication is obtained on the meter, the attenuation reading is noted. Then the
transmitter unit is placed in the required position on one side of
the wall and moved vertically and horizontally until a maximum signal
is obtained. The attenuator is adjusted again for center scale indication and the attenuator reading is noted. Since the attenuation
caused by a dry wall is found to be negligible, the difference between
the attenuation readings is a measure of the attenuation due to the
moisture in the wall. The moisture content is determined from calibration of moisture content against attenuation. In all measurements,
the cross-sectional area of the sample must be larger than the horn
aperture so as to minimize the leakage of energy around the material.
The sample should be levelled and ideally the horns should be in

contact with the material or be held in a constant relationship close to the surfaces. It is found, however, that with many materials a fixed horn spacing can be used with a constant cell height. In some cases the cell may consist of a waveguide directly coupled between transmitter and receiver. The small cross-sectional area enables the amount of material to be minimized when only small samples are available or if the water content is low it permits a greater thickness to be examined without requiring an excessive amount of material. In all measurements the percentage water is expressed in terms of the wet weight of the material. Using a constant wet weight of samples, the weight of water in the sample and hence the attenuation varies linearly with the percentage of water. In a measurement based on the difference of two attenuator readings the accuracy depends only on the attenuator. This is accurate to 0.1 dB. Thus if the sample is thick enough to give a change of attenuation of at least 3 dB for 1% change of water content, the error introduced by the instrument itself will not exceed 0.07%.

For building materials and structures a microwave moisture meter has several advantages over existing equipment and methods of moisture measurement. Because of the low attenuation of most basic dry materials, the meters measure only the water in a sample and nothing else. Further, no contact with the sample is necessary and so the method is nondestructive and cannot cause contamination. Measurements over comparatively large areas are possible, thereby minimizing sampling errors; in fact, samples need a minimum of preparation, the only consideration being the weight of a granular material or the thickness of a material of uniform density. The accuracy of measurements is not affected by the variable packing density of granular materials. Because the meter has the unique facility of being able to assess total moisture throughout the area covered by its horns, it has several uses in the building industry. It can be used to trace structural faults causing dampness in walls to test the efficiency of damp course, to locate woodwork concealed in walls and floors, and to assess the state of old buildings. In the latter case, a wall that appears on the surface to be thoroughly wet may be proved to be sound inside.

An interesting exercise with the meter has been carried out on the ancient structure of Trinity College, Dublin, where restoration work has been in progress. When originally built, the College's brick and rubble inner walls were tied at irregular intervals with timber ties of widely varying shapes and sizes. The outer walls were of solid granite facing stone and the intervening cavity was filled with rubble. Over the years, balks of timber had deteriorated through dry or wet rot, in some cases to the extent of having completely disappeared. Part of the reconstruction program involved the location of these timbers or the places where they had been so that they could be replaced by metal or concrete. Because of their random positions, however, there was no way short of a major structural examination of positively locating faulty timbers, or of ensuring that none had been missed. Systematic measurements with the AEI moisture meter over a section of the building solved the problem by revealing a definite change in power absorption wherever there was or had been a timber tie.

Among the industries interested in the moisture measurement of sand are iron and steel foundries, glass manufacturers, ready-mixed concrete manufacturers, refractory industries, and of course the building industry. While most interest will be in the process control applications and much of this centers around automatic monitoring systems, the less mechanized industries will find the microwave method, when applied to sampling techniques, has considerable appeal. In the foundry industry, the quality of moulds and consequently castings, is largely dependent on the water content of sand. The BCIRA (British Concrete Industries Research Association) and AEI have together developed a system of moisture measurement for automatic foundries which will shortly be available to industry. The system allows continuous monitoring and automatic control and has been proved by trial installation in a large modern factory. During trials, rapid automatic sampling of material on a conveyor belt enabled continuous monitoring of factory output to be recorded on a pen recorder and examination of the resultant chart revealed an accuracy of 0.25% over the total recording time. Sampling techniques result in even greater accuracy and under laboratory conditions, 0.1% accuracy can be achieved. The figure quoted can be regarded as applicable to measurement of all types of sand. Despite the complexity of the loss mechanism in microwave moisture measurement the results show that it is possible to obtain calibrations for materials which in many cases are linear in the region of interest. These results can be obtained in a simple manner with reasonable care and with regard to the fact that all materials are temperature sensitive.

1.8.6 Neutron-Scattering Method. The theory of the neutron-scattering method and its instrumentation have been discussed in Chap. VII and a number of publications [64-68].

The moisture-measuring method based on the slowing down of neutrons by hydrogen has many favorable features: The average value of moisture content can be obtained, the sampling error can be avoided, and it may be used as a signal for automatic control of the moisture content. These advantages have never been seen with conventional measuring methods. The radioactive method has been mainly used for soils and other similar types of materials, since in ordinary samples, the adsorption of neutrons and γ-rays may not be sufficient to give an accurate determination of its moisture content and density. The neutron-scattering method for measuring soil moisture fulfills the need in many types of studies to follow moisture changes in the solid without resorting to destructive sampling.

Most of the hydrogen found in the soil is contained in the soil water. A small amount is found in the mineral fraction, and this amount, being in chemical combination, should be relatively constant. Wetting or drying, as commonly understood, should not change the quantity of water in the mineral fraction. The most important source of hydrogen in soil, other than that of soil water itself, is the hydrogen in soil organic matter. According to Waksman [69], the

1 MOISTURE IN SOILS

hydrogen content of humus is about 5% of its weight. As the amount of hydrogen in water is about 11% of its weight, the amount of hydrogen in soil organic matter may be an appreciable part of the total hydrogen. But for practical agricultural work one must remember that soils containing much organic matter also contain large amounts of water. Earlier investigations [70-74] indicated that a universal relation or calibration curve could be obtained by employing the neutron method; later work done by some other researchers, however, indicates that this is not justified. Individual calibration relations or corrections must be obtained even though the deviations involved are of a minor nature.

The feasibility of an instrument for measuring moisture content of soils based on the principle of neutron scattering was first established by Belcher et al. [72]. Similarly, Spinks and co-workers [73] demonstrated the working of an experimental neutron moisture meter. Improvements in the technique and interpretation of the method were made by Sharpe [74], Van Bavel and co-workers [75], and Knight and Wright [76]. The fast neutron source usually consists of a mixture of polonium and beryllium or radium beryllium. The quantity of radium beryllium source usually employed varies from 20 50 mCi. A barium trifluorite tube is used as a thermal neutron detector, though earlier workers [77] used stable indium. Two types of probes known as (1) insertion type and (2) surface type are usually employed, which are shown in Fig. 12.21. The insertion type has a counting geometry of 4π while the surface type is limited to a counting geometry of 2π. Graphite is used as a reflector in the surface type of probe.

The radiations are counted by a slow neutron counter usually available commercially. This counter is usually cylindrical in shape and has a diameter of 5 to 6 cm and a length of 35 to 40 cm. The counter is placed immediately either on the top of the source of fast neutrons or aligned concentrically with it. These counters are usually operated by high voltages of the order of 3000 to 4000 V dc, which are highly stabilized to prevent damage to the counter. The

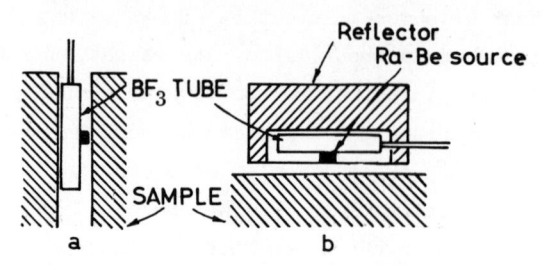

FIG. 12.21. Schematic arrangement of insertion-type neutron probe.

voltage stabilizer also eliminates variations in counting rate due to possible fluctuations in the line voltage of the power supply. The pulses obtained from the counter are fed through a linear amplifier into a count recorder which automatically records the number of radiations received by the detector. The count recorders are also usually commercially available and consist of a scale of 2 or 3 scalers. However, these recorders and scalers can be constructed in the laboratory as well. A block diagram of a typical neutron detection system employed by Stone et al. [78] is shown in Fig. 12.22. This shows a probe with cylindrical geometry.

The volume of soil which influences the counting rate of slow neutrons has a radius of about 20 cm at a water content of 0.2 g/cm^3. It is, therefore, evident that the measurement represents a weighted average property of this soil volume. The content of water in the soil close to the neutron source has a greater effect on the counting rate than has soil water further away. The variation of the moisture content of the soils and its relation with the depths below the soil surface has been theoretically and experimentally investigated by Holmes and Jenkinson [79]. According to their calculations, the following equation holds good:

$$\overline{W} = A + 2.83(\alpha - \beta) \tag{12.3}$$

where α and β are constants whose values depend on the change in gradient of the water content within the volume of soil measured.

1 MOISTURE IN SOILS   839

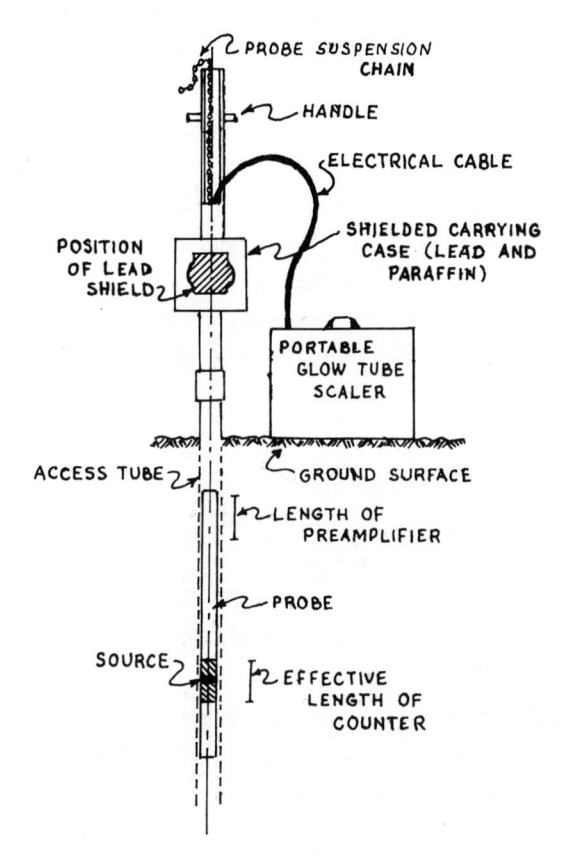

FIG. 12.22. Diagram of a neutron detection system.

The above equation has been experimentally verified by these researchers by measuring the moisture profile of a red-brown earth soil by the neutron method and subsequently determining the moisture content of the same soil by the gravimetric method using a soil-sampling technique.

The determination of field moisture by using the neutron-scattering method has also been made by Huet [80]. He has employed both the surface and depth types of moisture probes and finds that the linear response and the agreement between the computed and experimental curves obtained with the depth moisture probe are due to the fact that the detector is located wholly within the spheres of influence of materials with varying moisture content. The repeatability of counting rate was about 0.2 to 0.5% for total counts of approximately 40,000 to 50,000. Applications of the neutron moisture meter have been extended [81] to field problems in which measurements are required to be made by placing the equipment on the surface of the soil and similar materials like sand, sinter, ore mixture, etc. Unlike the earlier neutron meters which had their probes immersed inside the wet material, the probe in this case is laid on the surface of the material.

Huet [80] has also obtained comparative data of moisture content and dry density using surface and depth measurement techniques on a subgrade silty sand. It was observed that for the soils examined a mean difference in percentage of the dry weight of approximately 1% exists between the results obtained by radioactive and oven-drying methods. The possible explanation of the difference between the two methods may be obtained from a closer examination of the clay particles themselves, contained in the soil whose moisture contents have been determined. With the aid of the electron microscope and differential thermal analysis, it has been shown that clay minerals have a crystalline structure. They may be classified into three major groups, based on their lattice structures. These are kaolinite, montmorillonite, and illite. All of these groups have hydroxyl ions present in their lattices. Kaolinite has the largest number of hydroxyls, while montmorillonite and illite have about the same number each. These hydroxyls may provide sufficient hydrogen atoms to activate the neutron meter, causing it to indicate more moisture than is actually present. This may necessitate the application of a correction factor to the test results according to the soil type tested. If moisture variations in any one soil type can be successfully followed with the neutron meter, so that even if the exact water content is dependent on a correction factor, variations of moisture content can be detected without this factor. Although considerable work is still necessary to finalize the neutron

1 MOISTURE IN SOILS 839

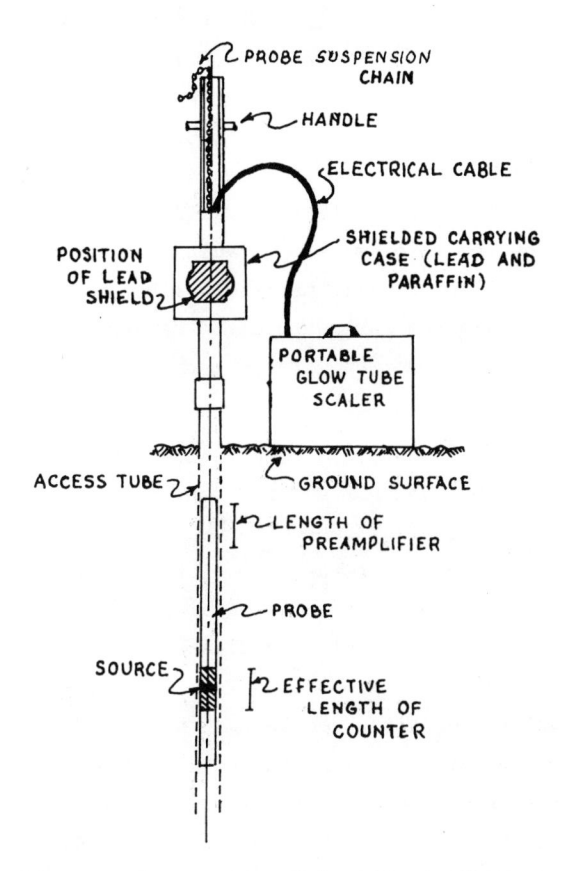

FIG. 12.22. Diagram of a neutron detection system.

The above equation has been experimentally verified by these researchers by measuring the moisture profile of a red-brown earth soil by the neutron method and subsequently determining the moisture content of the same soil by the gravimetric method using a soil-sampling technique.

The determination of field moisture by using the neutron-scattering method has also been made by Huet [80]. He has employed both the surface and depth types of moisture probes and finds that the linear response and the agreement between the computed and experimental curves obtained with the depth moisture probe are due to the fact that the detector is located wholly within the spheres of influence of materials with varying moisture content. The repeatability of counting rate was about 0.2 to 0.5% for total counts of approximately 40,000 to 50,000. Applications of the neutron moisture meter have been extended [81] to field problems in which measurements are required to be made by placing the equipment on the surface of the soil and similar materials like sand, sinter, ore mixture, etc. Unlike the earlier neutron meters which had their probes immersed inside the wet material, the probe in this case is laid on the surface of the material.

Huet [80] has also obtained comparative data of moisture content and dry density using surface and depth measurement techniques on a subgrade silty sand. It was observed that for the soils examined a mean difference in percentage of the dry weight of approximately 1% exists between the results obtained by radioactive and oven-drying methods. The possible explanation of the difference between the two methods may be obtained from a closer examination of the clay particles themselves, contained in the soil whose moisture contents have been determined. With the aid of the electron microscope and differential thermal analysis, it has been shown that clay minerals have a crystalline structure. They may be classified into three major groups, based on their lattice structures. These are kaolinite, montmorillonite, and illite. All of these groups have hydroxyl ions present in their lattices. Kaolinite has the largest number of hydroxyls, while montmorillonite and illite have about the same number each. These hydroxyls may provide sufficient hydrogen atoms to activate the neutron meter, causing it to indicate more moisture than is actually present. This may necessitate the application of a correction factor to the test results according to the soil type tested. If moisture variations in any one soil type can be successfully followed with the neutron meter, so that even if the exact water content is dependent on a correction factor, variations of moisture content can be detected without this factor. Although considerable work is still necessary to finalize the neutron

1 MOISTURE IN SOILS 841

moisture meter, the results obtained in various tests indicate that it is of considerable practical value in measuring moisture contents of soils and similar materials.

Belcher et al. [72] have compared the neutron-scattering method with the standard oven-dry technique by taking samples of soil every 6 in. during excavation of the holes for determination of moisture content by the standard oven-drying method. Density variations at each hole site are determined from an auxiliary 4-in. hole about 1 ft away, using the standard oil-displacement method of measuring volumes. Densities are determined for consecutive 6-in. layers. Two or three readings for moisture content at each depth were averaged to give a mean value. This is plotted in Fig. 12.23 against the average measured moisture content in the 6 in. above and below the source, since the zone of influence of the source is primarily a sphere of about 6 in. radius. Figure 12.23 shows the comparison of moisture-content determinations in a vertical hole. The neutron and γ-ray meter results are compared with actual oven-drying methods. Lane and co-workers [72] made 108 moisture-content determinations using the neutron meter, and the calibration curves were obtained from holes drilled on the campus of the University of Saskatchewan, Canada. These were compared with the values of water content obtained by the standard oven-drying method. Of these 65 were within 2%, 26 others were within 4%, and 17 more varied by more than 4%. All values refer to percentage water content.

The divergence in values obtained may be due to the presence of appreciable amounts of organic and inorganic materials such as ammonia nitrate, ammonia sulfate, iron, and chlorine, which would upset the calibration curve.

During manufacture or preparation of bulk materials such as foundry moulding sand, sinter ore mixture, constituents for concrete glass and ceramics, it is important to predetermine, to monitor, and if necessary to control the moisture content. Kosmowski [82] has measured the moisture content of such materials by using equipment

XII. MOISTURE IN SOILS, SANDS, CONCRETE, ETC.

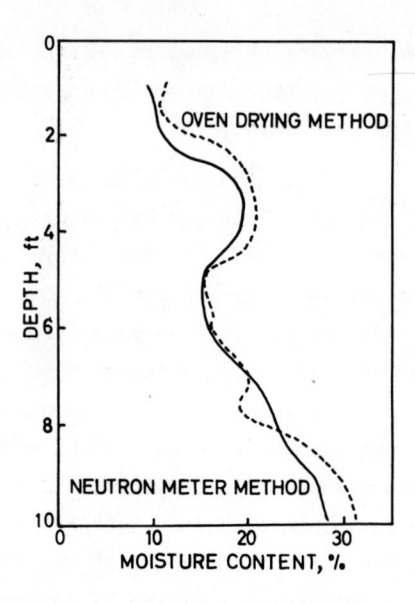

FIG. 12.23. Moisture profile comparison between standard oven-drying method and neutron meter.

consisting of a primary moisture element, an industrial rate meter, and a source holder with an americium-beryllium-neutron source having an activity of 100 mCi. For the purpose of measuring a moisture primary element, a pipe of 76 mm is built into the storage bin, bunker or tank. It contains at the lower end a number of neutron counter tubes, a preamplifier, and the source guide pipe. The neutron source is in the center of the counter tubes; pulses originating from the thermal neutrons are amplified, transformed, and fed into the industrial rate meter for indication. The equipment is calibrated after installation on account of the variable moderation effect of the dry part of the material under test and due to the existing conditions of installation, which are different from case to case. A definite minimum volume of bulk material is essential for the measuring process; the extreme lower limit lies at about 50 liters. If the measuring probe is uniformly surrounded by the material to be measured over a space of 60 cm diameter and 60 cm height, a linear relation is obtained between the measuring signal and the moisture content.

The use of the neutron and density meter presents a certain amount of radiation hazard. The danger from exposure is proportional to the distance between the source and the operator and to the length of time of exposure. Certain precautions to ensure the safety of the operation are therefore necessary. Although these are not very elaborate or expensive, nevertheless they are most important. An electromagnet mounted on a 3-ft length of $\frac{5}{16}$-in. rod, projecting from an ordinary flashlight case, facilitates the handling of the source. When not in use, the source should be stored in a lead castle. When it is in the ground, the operator should not stand directly over it. The operator should wear a film monitor on his person whenever he works with the meter. This film gives an indication of the exposure to γ rays and should be checked every week. The amount of neutron radiation which the human body can safely tolerate has been established. Keeping the allowable exposure down to one-third the tolerable amount for γ radiation (100 mR/week) should be safe. The intensity of the γ radiation from the Ra-Be neutron source at a distance of about 50 cm from the unshielded source would add up to a radiation dose of 300 mR in 40 hr. This is usually accepted to be the maximum permissible dose for a working week. The fast neutron intensity at about 20 cm from the unshielded source is the maximum permissible for a 40-hr/week exposure. It has been reported by Holmes and Turner [71] that a pocket dosimeter worn by operators of the equipment has measured on the average a weekly dose of about 0.03 mR of γ radiation for a usual amount of work with the apparatus.

From the discussion of the method based on the theory and the tests, the following conclusions can be drawn: The method appears valid from oven dryness to water saturation. The method yields the moisture content per unit volume directly and not moisture content per unit oven-dry weight. It requires but a single calibration for all mineral soils. However, the calibration depends on the depth of sample and other geometrical factors. It appears to be independent of soil temperature, texture, composition, compaction, and concentration of soil solution. It should be particularly useful in measuring soil moisture in the field in situ. This method for determining soil moisture is quite rapid, except at low moisture contents. At low moisture contents about 100 min would be required to obtain the needed 1000 counts in a usual apparatus. In some more suitable apparatuses the neutron source might be strong and more stable, so that correction for decay would not be required. In some apparatuses a counting rate meter is used instead of a counter with a watch.

The foregoing statements apply to mineral soils. Soils high in humus content, as mucks, present a special problem. The method should be adaptable to organic soils, however, if the hydrogen content in the organic matter when dry is constant. Some unusual soils contain boron in amounts up to 1%, and such a good neutron absorber would greatly reduce the counting rate. Chlorine has a quite high absorption cross section for slow neutrons, and sodium chloride is generally the main salt present in saline soils, which have an extensive occurrence in arid and semiarid regions. In saline soils for chloride concentration greater than 0.02 N a correction would have to be applied. Most of the hydrogen found in the soil is contained in the soil water. A small amount is found in the mineral fraction and this amount, being in chemical combination, should be relatively constant. Wetting or drying as commonly understood should not change the quantity of water in the mineral fraction.

It is also useful to study the effect of size of soil sample on counting rate. It was found by Gardner and Kirkham [68] that in water-saturated soil 1 ft away from the axis of the neutron source the effect would be larger. Small samples would be suitable according to the theory if a relatively smaller source and counter are used. But in most instances a large sample size should be a distinct advantage. For small samples, a collimated source would be desirable. With a collimated source the moisture content in long narrow glass or metal soil tubes could be followed without disturbing the soil in the tube.

1 MOISTURE IN SOILS   845

The depth of neutron source and counter in a soil auger hole will influence the counting rate, at least at depths comparable to the size of the source counter unit. But it would seem that the unit would not have to be placed in an auger hole. With proper calibration it would seem that a unit could be constructed for use at the soil surface. Design of a portable apparatus, suitable for field measurements, enhances the applicability of the technique to a wide range of problems. At present the application of the method is widespread in the United States, particularly in the fields of agriculture, water resources research, forestry, and more recently the construction industry. Cost and weight of equipment, lack of proper training of operators and of complete reliability of commercially available components are inhibiting a more rapid acceptance of the method. However, it is certain that with time these difficulties will be overcome. The health hazard associated with the use of radioactive materials does not seem to have prohibited the application in any appreciable way.

1.8.7 Tensiometric Method. The tensiometer is probably the easiest to install and the most rapidly read of all soil moisture-measuring equipments. Tension meters were probably most fully developed by Richards [59]. They are most useful for measuring the moisture content of tensions below approximately 0.9 atm. Such tensions will, on the average, correspond to a range in moisture content from slightly below field capacity to saturation. At the higher tensions found in drier soils, tensiometers become inoperative because air enters the system through the porous point. To determine the moisture content with a tensiometer, the relation between moisture tension and moisture content must be known. This relation may be found in the laboratory from a moisture-tension curve constructed by means of a pressure-membrane or porous-plate apparatus or by collecting soil samples in the area surrounding a tensiometer installation and relating the moisture content of the samples to the tensiometer reading obtained concurrently.

846   XII. MOISTURE IN SOILS, SANDS, CONCRETE, ETC.

A tensiometer consists of a porous point or cup (usually ceramic) connected through a tube to a pressure-measuring device. The system is filled with water and the water in the point or cup comes into equilibrium with the moisture in the surrounding soil. Water flows out of the point as the soil dries and creates greater tension, or back into the point as the soil becomes wetter and has less tension. These changes in pressure, or tension, are indicated on a measuring device, usually a Bourdon tube vacuum gauge or a mercury manometer. The tensiometer may also be attached to a pressure recorder or to an electronic pressure transducer to maintain a continuous record of tension changes. Tensiometers are available in lengths of 6 in. to 4 ft but probably could be manufactured in longer lengths if desired. Specially constructed tensiometers have been installed to depths of 15 ft. Multiple tensiometers for determining in tension data at several depths by use of a single probe were developed by Richards [59].

Tensiometers are affected by temperature. The temperature gradients between the porous point of the tensiometer and the soil may cause variations in the tension reading. The salt concentration in the soil or in the pore water seems to affect tensiometric methods less than electrical methods. Tensiometers exhibit considerable hysteresis effect; they tend to give a higher soil moisture tension during soil drying than during soil wetting. This effect is not too serious a disadvantage because the wetting cycle is usually rather short in comparison with the drying cycle. A serious disadvantage of the method is the time lag in respect to soil moisture changes. Tensiometers have exhibited lags of half an hour to many hours in indicating changes in tension caused in moisture content. In this connection it is interesting to note that recent studies by the hydrological laboratory of the US Geological Survey (Denver, Colorado) have shown that semipermeable plastic points provide much faster indication than ceramic points. Richards [59] noted that the vacuum-gauge type of instrument will generally provide an accuracy within 2% and the mercury manometer type is even more accurate. It

1 MOISTURE IN SOILS

may, however, be mentioned that tensiometers are not suitable for installation at depths greater than 20 ft.

A similar method for estimating moisture content consists of relating it to the force required to push an instrument through the soil. This type of instrument developed by Allyn and Work [83], which measures the force required to drive a pair of needles into a soil core, is known as an Availameter. They also reported a newly developed soil probe with which moisture-content estimation was possible within 0.5%. Equipments for measuring penetration resistance have also been developed. Penetration equipment has to be calibrated for each type of soil to obtain the relation between penetration resistance and moisture content. The method is very fast, although the equipment is difficult to use in gravelly or stony soils. A practical penetration equipment designed by the Waterways, Experiment Station Corps of Engineers (U.S.) consists of a pipe with a point at the bottom and a T-handle containing a pressure-indicating device at the top. Depth of penetration is, however, limited by the amount of force available. Recently Kazo [84] has developed and perfected a method for soil moisture determination on the basis of suction force with a tensiometer.

1.8.8 <u>Thermal Conductivity Method</u>. This method is based on the relation between moisture content and thermal conductivity of materials. If this relation is known for a certain material, the moisture content can be estimated from the thermal conductivity data. Though this method has been developed [84] for the determination of the moisture content of building structures it can also be applied for the estimation of moisture content of soils in bulk. By employing this method it is possible to determine the moisture distribution inside the material besides the measurement of the moisture content. The relationship between thermal conductivity and moisture content enables the moisture distribution to be determined as a function of the cross section of the structure, thus giving in many cases an indication of the source of moisture. The thermal conductivity of the material under investigation has been determined by Vos [85] by employing the nonsteady method developed by Blackwell [86], Woodside [87], and others. Vos [85] has mathematically derived the relationship between thermal conductivity and moisture content at different depths in a building structure and developed the experimental technique for its determination.

In order to measure the thermal conductivity at different distances or depths a thermal conductivity probe was designed and constructed [85]. The probe consists of a wire heater and a number of thermocouples which are built into a glass or polythene tube about 20 cm long. In order to make measurements in an existing wall, a hole is drilled in the wall and a glass tube is inserted. To ensure the fullest possible contact with the wall, the tube is coated with glue before insertion. The probe is pushed into the tube. By the use of this probe some difficulties were experienced by Vos [85] due to the temperature changes occurring in the structure which were of the same order as those that were to be measured. In order to overcome this drawback he developed a twin probe so that the cold junctions of the thermocouples in the one leg are put into a second leg. The distance between the two legs is so great that heat generated in the one leg during the experiment does not measurably penetrate to the other. The influence of extraneous nonsteady-state temperature effects, however, will be the same for both legs, an apparent steady-state thus being created as a result of which measurements can readily be made under these circumstances. The important thing in probe measurements is the determination of the temperature rise of the thermocouples as a function of time between 150 and 1000 sec. The part of the curve between switching on the heater and 150 sec is of no importance. After completion of the measurements the curves have to be put on simple logarithmic paper, after which thermal conductivity values can be determined from the slope. At first the temperature measurements were made with the aid of a hand-operated compensator. However, later on a recording method was developed which could make the temperature measurements entirely automatic.

1 MOISTURE IN SOILS

The measuring method described above gives the thermal conductivity value. For determining the moisture content from the thermal conductivity value the relationship between thermal conductivity and moisture content must be known for the material concerned. It is rather difficult to determine the thermal conductivity of moist materials, because a temperature gradient has to be established; as a consequence migration of moisture takes place, so that the original state is disturbed more or less. Experimental research in this field has been done on a large scale by Cammerer [88] and Jesperson [89]; Krischer [90] has made a theoretical study of this problem. This method has also been employed by Vos [85] for the investigation of the rain penetration as well as for the determination of moisture distribution within the thickness of the wall. The relationship between moisture content and the thickness of the wall is illustrated graphically in Fig. 12.24. Though this method does not give very accurate determination of the actual moisture content of building structures as would be expected by employing microwave absorption technique, it offers a very simple and comparatively less expensive method of approximately estimating the moisture content and moisture distribution within building structures which otherwise is difficult to be estimated by other conventional methods.

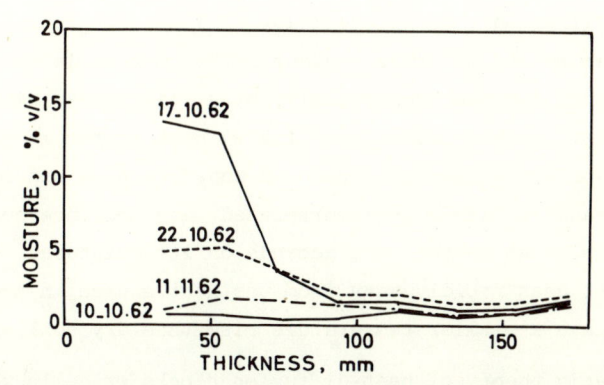

FIG. 12.24. Relationship between moisture content and thickness of wall.

Another method based on the thermal conductivity principle has been recently developed by Cambefort and Caron [91]. It consists in measuring the variation of heat loss in the soil. The measuring probe consists of a thermocouple and a heating wire protected by a glass tubing as shown in Fig. 12.25. Heating is provided by a 6-V battery and a rheostat which maintains a constant heating intensity at 300 or 600 mA. The temperature increments of the thermocouple are recorded every 30 sec during 10 min by a vacuum tube voltmeter which is accurate to one one-hundredth of a millivolt. The probe being fed with a constant current, its heating is proportional to the logarithm of time. This method is perfectly reproducible. The slope of the straight lines thus obtained depends on the moisture content of the soil or, if the soil is saturated, it depends on the soil property. In the case of nonsaturated sand, a slight discontinuity is observed when the sand-water-air mix is passed from the binary system sand-air (the quantity of water being too small to assure a continuous phase) to the ternary sand-water-air system. This discontinuity occurs for a moisture content corresponding to the water moistening the sand, viz., about 5%. The accuracy is very good in the sand and gravels. But these curves cannot be obtained in clays and muds. It seems that, owing to the presence of clay particles, the water does not behave freely as the water which flows in the sands and gravels.

Heat diffusion through the moist soil has also sometimes been used to measure its moisture content. This method is based on the principle that the heat conductivity of a soil varies with its moisture content. An electrically activated heat source is installed in the soil, and the temperature rise is measured by a sensitive temperature-measuring device and correlated with moisture content. The conduction of heat in the soil depends on its moisture. Wet soil will conduct heat rapidly away from the heat source in the cell and will thus have a smaller temperature rise than dry soil.

The basic theory of heat-diffusion blocks or cells was reported by Patten [92] as early as 1909. The design for a cell was suggested

1 MOISTURE IN SOILS   851

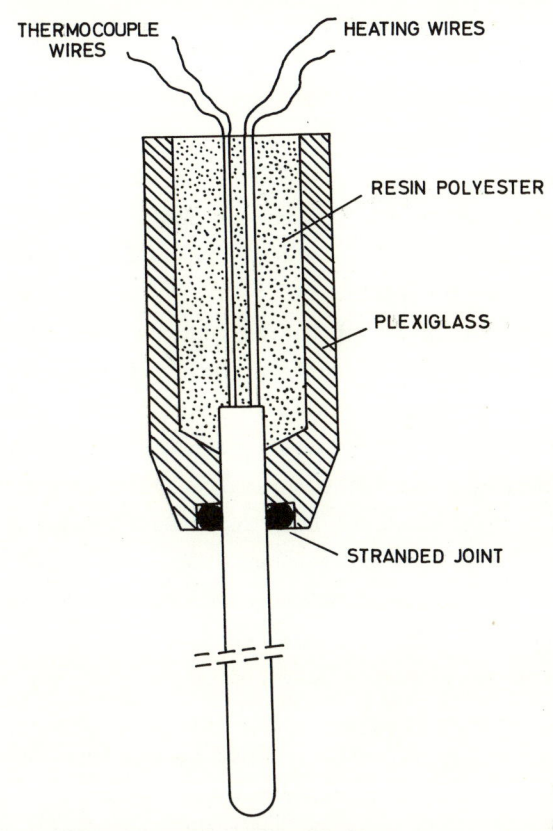

FIG. 12.25. Probe for thermal conductivity.

by Shaw and Baver [93] later in 1952. Kersten [94] tested several modifications of the design suggested by Shaw and Baver [93]. The heat-diffusion cells used to date can be classified into three general types as described by Aldous et al. [95]: (1) as a porous block type in which the electrical elements are imbedded in a porous medium, and (2) a direct-content type (thermal conductivity cell) in which the heater and the temperature-measuring elements are in contact with but separated by a portion of the soil being tested. The measurements to be made in this method are the recording of unbalance in a Wheatstone bridge resulting from the application of a heating current to a coil buried in the soil under investigation.

852   XII. MOISTURE IN SOILS, SANDS, CONCRETE, ETC.

Each soil has its own heat conductivity characteristics. Thus the curves that result from plotting data obtained by the use of either of the Shaw-Baver [93] circuits are affected by the heat conductivity of the soil in which the heater elements are buried. The result is that each soil in use must be calibrated with this electrical equipment before the instrument can be applied in the field. Salt concentrations from 100 to 1000 ppm do not affect the calibration data. Use of heat-diffusion cells has shown that the blocks are sensitive to minor variations in construction. The cells are unsatisfactory when used in soils at moisture contents above field capacity in high-shrinkage soils, intimate contact between the cell and the soil is lost as the moisture content decreases and erratic results are obtained until the shrinkage limit is reached. The porous-block type of cell has been reported as entirely unsatisfactory because consistent correlation between soil moisture and cell measurements could not be obtained under different soil conditions. The thermal-conductivity cell has been the most satisfactory of the three types but needs further development.

1.8.9 Hygrophotographic Method. This method consists of placing in contact with the soil a gelatine emulsion layer of silver and mercury iodide which absorbs the humidity at a speed which increases with the rise in moisture content of the soil, and which records the latter by a change in coloring proportionate to the amount of moisture absorbed. A means is thus provided of determining the amount of water, reckoned in micrograms, which a soil is capable of delivering to the hygrophotographic plate in terms of time and surface units.

This method gives measurement of moisture content present even in the minutest quantities and is specially suited for the determination of soil moisture following a protracted drought. This method has been employed by Sivadjian [96] for studying the effects of drought on the soil moisture and on the natural vegetation by recording and noting moisture at a depth of 12 in. or so. However, for an accurate determination of the microquantities of moisture it is necessary to supplement this technique by the densiometric scale. Since the recording requires a 45-min contact of the hygrophotographic plate with the soil, comparing this recording with the densiometric scale shows that this recording corresponds to a moisture content capable of yielding 0.18 µg water/min/mm^2 to the gelatin of the plate. The moisture content of the earth when the leaves of young wheat seedlings exhibit the first signs due to the drought was also investigated. It was shown that at that time, the earth could still yield, to the hygrophotographic plate at a depth of 20 mm, 3.25 µg water/mm^2/min.

This method is claimed to be highly sensitive and accurate though some doubts have been expressed that it is a dynamic measurement that depends upon the rate of supply of water to the plate. The accuracy of the measurement depends not only on the difference in water potential between the plate and the soil or plant tissue, but also upon the weight that water can move in the system. In this respect it would depend both on the water content and transfer coefficient or conductivity. For a detailed discussion of the technique and its application the reader is referred to the interesting publications of Sivadjian [96] who has also given a large number of pertinent references on the subject and compared this method with other well-known techniques.

2 MOISTURE IN CONCRETE AND SIMILAR MATERIALS

2.1 Introduction

The amount of moisture in concrete has a considerable bearing on its strength. The determination of water content in concrete is therefore important for predicting the durability of a concrete structure. The water permeability and water absorption are two other important factors. The tests used ordinarily for the determination of the comparative waterproof qualities of concrete, or better perhaps its comparative water tightness, are of two main types. In one case a direct determination of the water permeability is made; in the other, the water absorption of a dried concrete specimen is determined and this is then assumed to be proportional to the water permeability.

In determining this water absorption by the standard method, the concrete specimens are first dried at not less than 110°C to constant weight. They are then immersed in boiling water for 5 hr to drive out the air in the pores, and then cooled in water to a temperature of 10 to 15°C. They are then removed from water, superficially dried with towel or blotting paper and at once reweighed. The percentage of increase in weight when so saturated over that when thoroughly dried is termed the absorption of the specimen.

Here the underlying theory is evidently that a piece of concrete is made up of water-impervious masses, interspersed with pores or capillaries, and that the more pores or capillaries the concrete has the greater will be its water absorption and the less its water tightness; or in other words, that the chief reason for the permeability of concrete is its porosity and that the water absorption of the dried specimen is the correct measure of this porosity. In determining the water permeability directly two general methods are available. In one, exemplified by the standard hydrostatic test for cement pipe, the specimens are subjected to an increased water pressure until the water is forced through, and the pressure at which this occurs is taken as a measure of the permeability. In the other method, the cement specimens are subjected to a constant water pressure and the weight of water forced through a unit area in unit time is determined, and becomes the measure of the permeability of the specimens. Experimental results, however, show [92] that there is in general no direct relation between water permeability and the water absorption

2 MOISTURE IN CONCRETE AND SIMILAR MATERIALS

and that the water permeability of concrete cannot be correctly determined by measuring its water absorption.

2.2 Methods of Measurement

2.2.1 <u>Neutron-Scattering Method</u>. After the successful application of the neutron-scattering method of moisture determination in soils Pawliw and Spinks [77] modified the method to be suitable for nondestructive determination of water in concrete. The Pawliw and Spinks [77] apparatus (Fig. 12.26) designed for concrete consisted of 50 mCi Ra-Be neutron source kept in a lead container. The detector used was a BF_3 tube. It has a copper cathode 0.75 in. in diameter, 0.05 in. thick with a sensitive length of 4 in. The counter tube is filled with B^{10} and enriched BF_3 to a pressure of 60 cm Hg. Since the capture cross section for 0.1 B^{10} is 39,990 barns for thermal neutrons and less than a barn for MeV neutrons, such a tube forms an excellent detector for slow neutrons and gives instantaneous reading.

The neutron source is held by a clamp as close as possible to the BF_3 counter since, as pointed out earlier, both the theoretical considerations and experimental observations indicate that maximum sensitivity of the soil moisture meter is obtained when the fast neutron source is as close as possible to the slow neutron detector. Furthermore, the effect on the counting rate of neutrons slowed down by collisions with nuclei other than hydrogen can be eliminated by placing the slow neutron counter as near as possible to the source

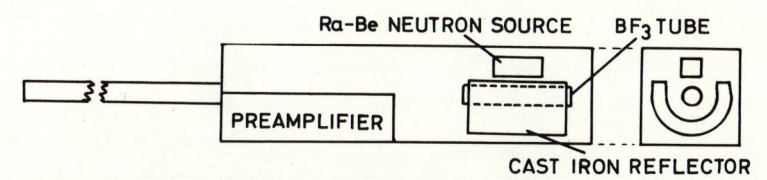

FIG. 12.26. Neutron moisture meter for concrete.

of fast neutrons. In this position the counting rate is almost
entirely a function of the amount of hydrogen present in the surrounding medium. In the case of soils, to indicate water content
the moisture meter is placed in a hole in the soil. Thus the meter
is almost completely surrounded by the neutron-thermalizing material.
The concrete moisture meter, on the other hand, cannot be placed in
a hole in the concrete surface. Consequently, for this geometry
the effective neutron-thermalizing volume of a concrete member will
be considerably less than it is for a soil body and the sensitivity
will be correspondingly less.

As a means of counteracting this decrease in sensitivity, a U-shaped neutron reflector was placed over the counter (Fig. 12.26) and source in such a manner as to reflect some of the neutrons which would otherwise have escaped, into the concrete. To prevent appreciable neutron capture and thermalizing effect, the reflector material should have a small capture cross section and low neutron-thermalizing capacity. Such a reflector would merely scatter the fast neutrons and increase the number of these neutrons entering the concrete. Furthermore, slow neutrons returning to the surface by back diffusion might also be directed toward the counter by scattering within the reflector. A number of steel reflectors of varying thickness can be placed over the BF_3 counter. Experiments indicate that a U-shaped neutron reflector will appreciably increase the sensitivity of the concrete moisture meter. Thus the effect of a cast iron reflector is to increase the thermal neutron count by a factor of 2.35.

The external structure of the meter is so designed as to enable an operator to handle the meter during short intervals of time without exposing himself to excessive radiations from the source. A high-gain, four-stage amplifier is employed to receive the small BF_3 counter pulses and amplify them sufficiently to pass through the long connector cable and trip the scaling mechanism. To correct the fluctuations in the background count and slight variation in the counter and/or preamplifier voltage, a standard paraffin neutron

2 MOISTURE IN CONCRETE AND SIMILAR MATERIALS

thermalizer has been constructed [77]. This enables the operator to compare all observations with the response from a constant neutron-thermalizing medium. All observations are recorded as the relative thermal neutron count, which is given as:

$$N_r = \frac{\text{observed count - background count}}{\text{standard paraffin count - background count}} \qquad (12.4)$$

The background count and the standard paraffin count were determined immediately after the required observation was recorded.

The effective neutron-thermalizing volume of a substance is dependent on the water content of the substance. It is therefore necessary to determine edge and depth effects. The investigation of these effects is facilitated by using synthetic mixtures of sand and sugar in place of concrete blocks. Sugar may be regarded as 57.9% water and 42.1% carbon and a mixture of sand and sugar of any given concentration has a corresponding equivalent water content. The neutron-thermalizing effect of the mixture depends on this equivalent water content. Edge and depth effects can then be investigated using cylindrical vessels of various radii and depths, filled with the mixtures and with water. For calibration, two concrete cylinders are usually made, each with a diameter of 18 in. and depths of 8.3 and 8.75 in., respectively. Each cylinder consists of three slices of equal thickness. This simplifies the handling of the concrete specimens and makes the removal of the water from the concrete somewhat more uniform.

In the experiments of Pawliw and Spinks [77] each concrete specimen was prepared by mixing 23 lb Portland cement, 74 lb lightweight aggregate, and 21 lb water. After 48 hr the forms were removed and the concrete was placed in water for several days. These concrete cylinders were then used to calibrate the concrete moisture meter. The water was removed from the concrete by heating the concrete in an oven at 130°F. This drying process was carried out in a stepwise manner, and after each drying period the concrete was

858   XII. MOISTURE IN SOILS, SANDS, CONCRETE, ETC.

weighed and a relative thermal neutron count was determined using the concrete moisture meter. Finally, when most of the water had been removed, the concrete cylinders were crushed with a jaw crusher and representative samples removed for further drying. In order to remove the total water including the water chemically combined to form hydration products, these samples were heated to 1000°C. The calibration curve for concrete together with results from the sand-sugar experiments are shown in Fig. 12.27.

The theoretical method of calibration of a neutron moisture meter given by Holmes [70-71] has been applied by Pawliw and Spinks [77] to the case of concrete. The relative thermal neutron count N_1 indicated by the moisture meter is given as follows:

$$N_1 = NWF \qquad (12.5)$$

where N is the relative thermal neutron count indicated by the moisture meter for water; W is the water content of the soil; and

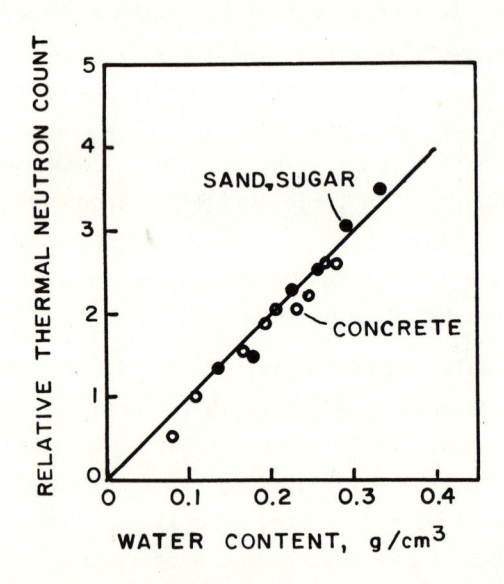

FIG. 12.27. Relative thermal count versus water content for concrete and sand-sugar.

F is the factor whose value depends on the slowing-down length of neutrons in water, diffusion length of neutrons in water, and distance between source and point where the thermal neutron flux is measured.

The value of γ, which is the distance between source and point where the thermal neutron flux is measured, cannot be found directly. It can, however, be calculated from experimental data obtained from a sample of known water content. In other words, knowing N_1, N, W, and F, a value for γ can be found by trial and error to give the required F; γ depends on the geometry of the instrument and is constant for different values of W. This method is found suitable for the concrete moisture meter. The value of γ for the concrete moisture meter was found to be 3 cm; F remained very nearly constant for all values of W being considered. This made the theoretical calibration curve, calculated from Eq. (12.5), a straight line. This curve is plotted in Fig. 12.27 together with the experimental plots. This shows quite good agreement between theory and experimental results.

From what has been described, it is clear that it is quite feasible to construct a portable moisture meter in the laboratory which will give a rapid and nondestructive measurement of the moisture content of a concrete member. Care would have to be taken in interpreting the readings if the concrete member contained any appreciable amount of steel reinforcing or if the moisture distribution were nonuniform. As a matter of fact, quite a few commercial neutron moisture meters have been developed and manufactured in the USA, the UK and USSR which are quite portable and convenient for routine observations of moisture contents in field operations.

2.2.2 Electric or Electronic Method. Besides the neutron-scattering method described in the earlier sections, a few investigators have developed electrical resistance and dielectric types of moisture meters for studying the moisture content of concrete and similar materials. Ghosh and Khanna [97] have developed an electrical resistance method for estimation of moisture differential in concrete pavement slabs. They have designed electric resistance moisture cells made of cement mortar which were found to give greater sensitivity to moisture than plaster of Paris blocks whose constructional features have been described under the electric or electronic method of moisture measurement given earlier in the chapter. The electrodes of the cement mortar resistance cells were made of insulated copper wire $\frac{1}{30}$ in. thick and 0.78 in. square and copper plates were soldered at the end. These electrodes were embedded in 1 to 1.5-in. sand proportions of the concrete mix whose moisture content in hardened state was to be measured. The resistance was measured directly with the help of a universal Wheatstone-type measuring bridge which could measure up to a maximum of 10 Mohms. Details of the apparatus are given in Fig. 12.28. First, the electrodes were kept fixed in position at the center of the mold and the gap between the two plates was filled with mortar, followed by complete filling of the mold by mortar. After 24 hr, the cells were removed from the mold and stored in damp atmosphere for 28 days.

In the experiments conducted by Ghosh and Khanna [97] mortar samples within 24 hr of destripping from the mold were insulated with polyethylene at the sides and bottom, and asphalt and paraffin at the top to prevent any moisture changes occurring. The resistance of the cells was measured with the four-electrode system. The calibration was done within the range of the maximum degree of saturation obtained

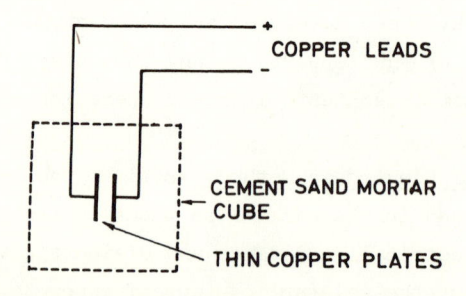

FIG. 12.28. Cement mortar moisture gauge.

by simple soaking in water and the minimum by drying in air at room temperature. This was done in accordance with the findings of the Building Research Station, Watford (UK) which showed much less hysteresis under the conditions than in the case when calibration was begun from a vacuum saturated state. Although the above procedure of calibration and the matching of pore structure and pore size distribution by adopting cement mortar as the base for the cell for the purpose under consideration reduce the degree of hysteresis considerably, it could not be eliminated completely. It was found by these investigators that the hysteresis loss in the case of plaster of Paris cells is less than that in cement mortar cells which were used by them. This they attributed to the fact that the cement mortar is less homogeneous than the plaster of Paris. The pore structure and pore size distribution in plaster of Paris are such that the effect of drying and wetting reaches the center of the cell very rapidly, which was not the case with the cement-sand mortar blocks. Figure 12.29 shows the relationship between the moisture content and cell resistance for various water-cement ratios and proportions of mortar mixes. These indicate that a higher water-cement ratio offers lower resistance to conductivity provided the water-cement ratio is such that the mix can be well compacted. Where the water-cement ratio was very low, say 0.35, the compaction was not perfect and higher porosity in the mix results in lower resistance to conductivity at different moisture contents. As a result of laboratory and field studies carried out by Ghosh and Khanna [97] it was found that a sand-cement-mortar moisture cell may be used for measuring moisture content in other porous building materials in studying problems caused by the moisture. The measurement of moisture variations has special significance in soil studies in connection with foundation problems because bearing capacity of many soils varies greatly with the moisture content. These cells may also be used as practical tools for soil moisture measurement in irrigated fields. In Japan resistance-type cells have been used to measure moisture content in walls and ceiling plasters before any painting or distemper work is done on them, especially during high-humidity summer months.

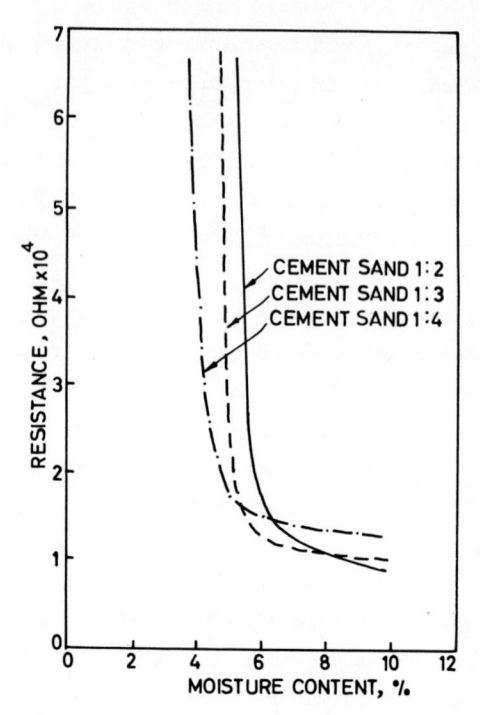

FIG. 12.29. Resistance of moisture cells made of cement mortar with different W/C ratios.

Szuk [98] has devised a simple apparatus for the measurement of surface moisture of building materials and has applied it for the study of surface moisture-migration phenomena. The problem of surface moisture of buildings and building materials has a special significance since floor and wall claddings and wall paints based on synthetic materials are in use. When applying wall paints and claddings made of synthetic materials, moisture content of the support or, better, the humidity conditions of the surface to be sealed or coated is to be determined to avoid risk of ulterior deteriorations. The degree of humidity of the surface to be painted or coated is a decisive factor in the durability of coatings. It is often found that plastic wall paint blisters or peels off, bonded plastic wall tile or sheets leave the wall, and bonded floor cladding lifts off or, what is worse, blisters.

The apparatus used by Szuk [98] is based on the ohmic resistance of the building material, which varies with the surface moisture of the material. In order to eliminate the error caused by the contact resistance between the electrodes and the materials a special sensing unit has been developed with electrodes made of an elastic foam with an electrical conducting surface. The measurement of the resistance with a sensing unit is done by pressing against the surface of the wall and noting the ohmic resistance from the ohmmeter. This resistance with a calibration chart for the material to be tested gives directly the moisture content. A single measurement lasts about 5 sec. The instrument can measure moisture contents from 0 to 15%.

Dielectric measurements on Portland cement paste have been carried out by De Loor [99] at frequencies of 10^5 and 10^7 Hz and at microwave frequencies of 5500, 9700, and 16,500 MHz. The Portland cement studied by him had the following composition: C_3S 60.1%, C_2S 14.4%, C_3A 10.1%, C_4AF 8.6%, in which C stands for CaO, S for SiO_2, A for Al_2O_3, and F for Fe_2O_3. Three types of samples were prepared, the first having a wc ratio of 0.31 was prepared in a closed mold and cured under water for 1 month. The second series having a wc ratio of 0.24 was made in open molds and cured in the laboratory atmosphere for 3 weeks at 50% rh and a temperature of 23°C for 5 months under distilled water. The same procedure was applied for the third series with a wc ratio of 0.24 at 34% rh and 20°C temperature, which was only measured during the first week of hardening. To obtain dry samples, they were put into an oven at 100°C for 14 days and further dried in a desiccator over P_2O_5 under vacuum. It was found that there is a linear relationship between the dielectric constant and the moisture content percentage on all the frequencies employed by him for the measurement. From these observations it is apparent that based on these properties the dielectric meter can be employed as a moisture meter with proper calibration. Moisture measurement of building materials such as sand, cement, clays, minerals, and lime muds has been done by Legutin et al. [100] by the method based on the attenuation of microwaves and ultrasonics. They also developed an automatic microwave moisture meter using an automatic compensating bridge for measurement and control of high moisture contents (up to from 30 to 45%).

3 MOISTURE IN SILICA AND SILICATES

3.1 Introduction

Silica and silicates occupy a unique position among natural and artificial absorbents in the inorganic world. There are almost infinite variations in their absorbing properties, and due to the method of preparation even a slight change in the technique produces a profound change in the absorption. The only explanation of this fact is that the vapor absorption is due to capillary condensation as the size distribution of capillaries would admit of infinite variations. Moisture absorption by silica and silicates prepared in a variety of ways thus offers a fruitful line of investigation in elucidating the mechanism of moisture absorption by capillary systems in general. Moisture absorption by various activated silica samples prepared in different ways has been studied by Puri [2]. The following general conclusions can be drawn from the data obtained:

  1. There is a gradual increase in the absorptive capacity as the metallic ion is removed.

  2. In some cases there seems to be an optimum beyond which, if the metallic ion is removed, the absorptive capacity tends to fall.

  3. Ni, Cu, and Ba on the whole show lower absorptive capacity when they are removed from their corresponding silicates than do silicates of Al, Cr, CO, and Fe.

3 MOISTURE IN SILICA AND SILICATES

3.2 Methods of Measurement

3.2.1 Penfield Method. In the usual method for the determination of water in silicate rocks, the powdered sample is ignited either by itself or with a flux, and subsequent measurement is made of the expelled water. The expelled water can be measured in several ways. It can be absorbed in a desiccant in a preweighed tube, condensed, and determined by measurement of volume, or measured as in the method of Penfield [101], which is more widely used than any other method. In the Penfield method the upper part of the glass tube, containing the condensed water expelled from the sample powder, is separated by fusion from the lower part of the glass tube containing the powder. The part containing the water must be allowed to reach thermal and moisture equilibrium with the laboratory air before weighing. A second, time-consuming heating and equilibration step is necessary in getting the final or "dry tube" weight to subtract from the initial weight to obtain the amount of water in the sample. The Penfield method usually requires from 30 to 40 min for a single determination, and the value of moisture content obtained for total water in a rock analysis includes water of crystallization, water held uncombined in the grains or on their surfaces, and water formed as a result of heating from hydrogen or hydroxyl groups present in regular atomic arrangement in the molecular or crystal structure.

3.2.2 Loss-on-Ignition Method or Rapid Method. The value of moisture content can also be obtained by determination of the loss on ignition. Values so obtained vary in accuracy from good to poor depending upon the composition of the samples. Shapiro and Brannock [102] describe the ignition-loss method of water determination in rocks. For samples in which carbonate and ferrous iron concentrations are low, values based on the ignition-loss determination are entirely adequate. When the samples contain appreciable carbonate or appreciable ferrous iron or both, direct determination of water is desirable. This method is a modification of the Penfield procedure; it requires less than 10 min/sample and was developed to

866   XII. MOISTURE IN SOILS, SANDS, CONCRETE, ETC.

meet the need for a rapid method for the determination. It consists of expelling the water from the sample in a test tube, absorbing the water on a preweighed strip of filter paper placed inside the tube, and subsequently reweighing the paper. The difference in the weight of the paper before and after absorption of the water is then used, after correction for water lost in handling, to calculate the percentage of water in the sample. The apparatus is shown in Fig. 12.30. For details of the procedure, the reader is referred to the original publication [102].

Results obtained by the rapid method for total water in eight rock samples by two analysts are compared in Table 12.1 with results obtained by the conventional Penfield procedure. The values are in close agreement. A collaborative study of silicate rock analysis reports the results of the water determinations by the Penfield

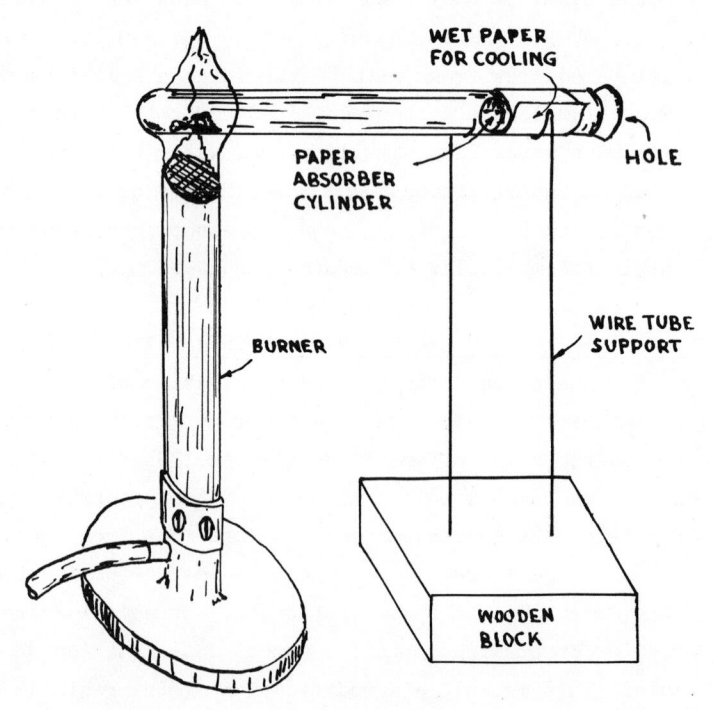

FIG. 12.30. Apparatus for determination of water in rock silicates.

REFERENCES

method and some of its modifications. The reported values varied from 0.3 to 0.4% at the 0.5% level of water. The rapid method gives values well within the range indicated by this collaborative study.

TABLE 12.1

Comparison of Results
by Rapid and Penfield Methods

Sample	Water (%) Rapid	Water (%) Penfield
1	0.58	0.54
2	0.40	0.35
3	1.00	1.00
4	2.1	2.1
5	2.3	2.4
6	2.2	2.3
7	7.8	7.8
8	13.4	13.4

REFERENCES

1. G. J. Bouyoucos, Soil Sci., 67, 319 (1949); 76, 447 (1953); Agronomy J., 42, 104 (1950); 43, 508 (1951); 44, 311 (1952); Michigan Agr. Expt. Sta. Quart. Bull., 37, 132 (1954); Humidity and Moisture, vol. 4, Reinhold, New York, 1965, pp. 105-111.

2. A. N. Puri, Soils, Their Physics and Chemistry, Reinhold, New York, 1949.

3. W. P. M. Black, D. Croney, and J. C. Jacobs, Field Studies of the Movement of Soil Moisture, Paper No. 41, Road Research Laboratory, Department of Scientific and Industrial Research, U.K.

XII. MOISTURE IN SOILS, SANDS, CONCRETE, ETC.

4. D. Croney, Geotechnic, London, 3, 1-16 (1952).

5. G. B. Bodman and E. A. Colman, Proc. Soil Sci. Amer., 8, 116 (1942).

6. W. Gardner and J. A. Widtsoe, Soil Sci., 11, 215-233 (1921).

7. E. C. Childs, J. Agr. Sci., 114-141, 527-545, I and II (1936).

8. N. A. Ostashev, The Law of Distribution of Moisture in Soils and Methods for the Study of the Same, Internal Conf. Soil Mechanics and Found. Eng. Proc. I (Sect. K), 1936, pp. 227-29.

9. D. Kirkham and C. L. Peng, Soil Sci., Jan.-June, 1949, p. 29.

10. P. T. John, Ind. J. Meteorology Geophysics, 13, 71 (1962).

11. R. K. Schofield, The P.F. of Water in Soil in Relation to Highway Design and Performance of Water and Its Conduction in Soils, S. R. 40 High Way Research Board, 1958, pp. 226-53.

12. A. I. Johnson, Methods of Measuring Soil Moisture in the Field, Contributions to the Hydrology of the United States Geological Survey Water Supply Paper 1619.

13. M. B. Russel, Soil Sci. Soc. Amer. Proc., 14, 73 (1950).

14. M. A. Whitney, Instructions for Taking Samples of Soil for Moisture Determination, U. S. Dept. Agr., Div. Soil Circular No. 2, 1894; also Methods of Mechanical Analysis of Soils and of the Determination of the Amount of Moisture in Soils in the Field, U. S. Dept. Agr. Div., Soils Bull. No. 4, 1929, p. 24.

15. F. J. Veihmeyer, Soil Sci., 27, 147 (1929).

16. H. A. Noyes and J. F. Trost, Assoc. Offic. Anal. Chem. J., 4, 95 (1920).

17. J. F. Reed and J. A. Rigney, Amer. Soc. Agronomy J., 39, 26 (1947).

18. K. G. Reinhart, Relation of Soil Bulk Density to Moisture Content as It Affects Soil Moisture Records, U. S. Dept. Agr. Southern Forest Expt. Sta. Occasional Paper, 135, 12 (1954).

19. U. S. Bureau of Plant Industry, Soils and Agriculture Engineering, Soil Survey Manual; U. S. Dept. Agr. Handbook 18, 1951, p. 327.

20. C. H. M. Van Bavel and M. J. Gilbert, Discussion of a Method for Approximating the Water Content of Soils, Am. Geophys. Union Trans., 35, 168 (1954).

REFERENCES

21. M. Stephenson, *J. Text. Inst.*, 29, T297 (1938).
22. G. F. Davidson and S. A. Shorter, *J. Text. Inst.*, 21, T165 (1930).
23. J. H. Thompson, *Ind. Chem.*, 34, 451 (1958).
24. H. S. Booth and H. L. McIntyre, *Ind. Eng. Chem. (Anal. Ed.)*, 2, 12 (1930).
25. R. N. Evans, J. E. Devanport, and A. Revukas, *Ind. Eng. Chem. (Anal. Ed.)*, 11, 553 (1939).
26. IS — 2720 (Part I) — 1966 Indian Standard Methods of Test for Soils — Part I, *Preparation of Dry Soil Samples for Various Tests*, UDC 624-131.4; 543-05; also IS — 2720 (Part II) — 1964, Indian Standard Methods for Soils — Part II, *Determination of Moisture Content*, UDC 624-131.43; 543-812.
27. G. J. Bouyoucos and A. H. Mick, *Michigan Agr. Expt. Sta. Tech. Bull. 170*, 1940; *Soil Sci.*, 63, 455-65 (1947).
28. E. A. Colman, *Am. Geophys. Union Trans.*, 27, 847 (1946).
29. R. E. Youker and F. R. Dreibelbis, *Am. Geophys. Union Trans.*, 32, 447 (1951).
30. B. L. Korty and H. Kohnke, *Soil Sci. Soc. Amer. Proc.*, 17, 307 (1953).
31. E. A. Colman and T. M. Hendrix, *Soil Sci.*, 67, 425 (1949).
32. H. A. Weaver and V. C. Jamison, *Agronomy J.*, 43, 602 (1951).
33. O. J. Kelley, *Soil Sci.*, 58, 433-40 (1944).
34. A. B. C. Anderson, *Soil Sci.*, 56, 29-41 (1943).
35. A. B. C. Anderson, N. E. Edlefsen, and W. B. Marcum, *Soil Sci.*, 54, 275-79 (1942).
36. A. B. C. Anderson and N. E. Edlefsen, *Soil Sci.*, 54, 35-46 (1942).
37. A. B. C. Anderson and N. E. Edlefsen, *Soil Sci.*, 53, 413-28 (1942).
38. E. C. Childs, *Soil Sci.*, 46, 95-106 (1938).
39. A. Pande, *Instrum. Pract. (London)*, 15, 432 (1961).

40. A. Pande, *Instrum. Pract. (London)*, 19, 650 (1965).

41. H. Fricke and H. J. Curtis, *J. Phys. Chem.*, 41, 729-45 (1937).

42. H. Fricke and A. Parts, *J. Phys. Chem.*, 42, 1171-85 (1938).

43. H. Fricke and L. E. Jacobson, *J. Phys. Chem.*, 43, 781-96 (1939).

44. W. H. McCorkle, *Texas Agr. Exp. Sta. Bull.* 426, 1931.

45. W. G. Smiley and A. K. Smith, *J. Amer. Chem. Soc.*, 64, 624-28 (1942).

46. L. J. Briggs, Electrical Instruments for Determining the Moisture Temperature and Soluble Salt Content of Soils, *U. S. Dept. Agr. Div. Soils Bull.*, 15, 1-35 (1899).

47. J. Mathews, *J. Agr. Eng. Res.*, 8, 17-30 (1963).

48. G. Lejeune and G. Arnould, *C. R. Acad. Sci., Paris*, 246, 1217-19 (1958).

49. A. F. Kugaevskii, *Measuring the Complex Permittivity and Permeability of Materials with a Q Meter*, English Trans. Instrum. Exper. Tech. (USA) 364-7 (1962), published November, S.A.B., 8257 (1963).

50. D. J. Millard, *Brit. J. Appl. Phys.*, 4, 84-7 (1953).

51. T. S. McLeod and A. E. Yallup, Proc. I.E.E., 108B, 449 (1961).

52. A. M. Thomas, *An Experimental and Theoretical Study of in situ Measurement of Moisture in Soil and Similar Substances by 'Fringe' Capacitance*, Electrical Research Associates U.K. Report No. 5032, 1965, pp. 1-41.

53. A. R. Von Hippel, *Dielectric Materials and Their Applications*, Chapman and Hall Ltd., London, 1964.

54. C. V. De Plater, *Soil Sci.*, 80, 391-95 (1955).

55. A. F. Labedeff, Proc. First Int. Confr., Soil Sci., Washington, D.C., 1952.

56. M. Rocha, Proc. IVth Int. Conf. Soil Mech. and Foundation Eng., London, 1957.

57. H. Fukuda, *Soil Sci.*, 81, 2 (1956).

REFERENCES

58. D. De Costro, *Humidity and Moisture*, vol. 4, Reinhold, New York, 1965, p. 7.

59. L. A. Richards, *A Thermocouple Psychrometer for Measuring the Relative Vapour Pressure of Water in Liquids or Porous Materials*, Proc. Int. Symp. on Humidity and Moisture, Washington, D.C., USA, 1963.

60. R. P. Teele and S. Schuhmann, National Bureau of Standards Research Paper RP 1195, *J. Res.*, 22, 431-39 (1939).

61. L. A. Richards and G. Ogata, *Science*, 128, 1089-90 (1958).

62. J. B. Hasted and M. A. Shah, *Brit. J. Appl. Phys.*, 15, 825 (1964).

63. H. B. Taylor, *AEI Eng.*, Jan./Feb., 3 (1965).

64. J. Corck, *Radioactivity and Nuclear Physics*, D. Van Nostrand, New York, 1947.

65. R. K. Adair, *Rev. Mod. Phys.*, 22, 249 (1950).

66. R. E. Marshak, *Rev. Mod. Phys.*, 19, 185-238 (1947).

67. J. W. Holmes, *Aust. J. Appl. Sci.*, 7 (1), 45 (1945).

68. W. Gardner and D. Kirkham, *Soil Sci.*, 5, 391 (1952).

69. S. A. Waksman, *Humus, Ed.*, 2, 49 (1938).

70. J. W. Holmes, Aus. Atom. Energy Comm. Symp., 1958.

71. J. W. Holmes and K. G. Turner, *J. Agr. Eng. Res.*, 3, 199 (1958).

72. D. J. Belcher, T. R. Cuykendall, and H. S. Sack, *The Measurement of Soil Moisture and Density by Neutron and Gamma Ray Scattering*, Tech. Rep. 127, U.S. Civ. Aero. Admin., 1950.

73. J. W. T. Spinks, D. A. Lane, and B. B. Torchinsky, Symp. on the use of radioisotopes in soil mechanics. Special technical publication by the American Society for Testing Materials, 1952; *Can. J. Technol.*, 29, 371 (1951).

74. J. Sharpe, Brit. *J. Appl. Phys.*, 4, 93 (1953).

75. C. H. M. Van Bavel, N. Underwood, and R. W. Swanson, *Soil Sci.*, 82 (1), 29 (1956).

76. A. H. Knight and T. W. Wright, Soil Moisture Determination by Neutron Scattering, *Radioisotope Conf., Oxford*, 2, 111 (1954).

77. J. Pawliw and J. W. T. Spinks, Can. J. Technol., 34, 503 (1957).

78. J. F. Stone, R. H. Shaw, and D. Kirkham, Soil Sci. Soc. Amer. Proc., 6, 435-38 (1960).

79. J. W. Holmes and A. F. Jenkinson, J. Agr. Eng. Res., 2, 100-109 (1959).

80. J. Huet, Humidity and Moisture, vol. 4, Reinhold, New York, 1965, p. 185.

81. Y. Miyashita, Humidity Moisture, 4, 195 (1965).

82. A. Kosmowski, Instrum. Prac., 20, 669-72 (1966).

83. R. B. Allyn and R. A. Work, Soil Sci., 51, 307-19, 391-406 (1941).

84. B. Kazo, Proc. Imeko Symposium on Moisture Measurement, Hungary, 1971, p. 241.

85. B. H. Vos, Humidity Moisture, 4, 35-47 (1965).

86. J. H. Blackwell, J. Appl. Phys., 25, 2 (1954).

87. W. Woodside, ASHRAE J., Sept. (1958).

88. J. S. Cammerer, Warme- Kaltetechnik, 41, 126 (1939).

89. H. B. Jesperson, J. Inst. Heating Ventilating Engrs., 21, 157 (1953).

90. O. Krischer, Die Wissenschaftliche Grundlagen der Trocknungstechnik, Springer Verlag, Berlin, 1956.

91. H. Cambefort and C. Caron, Humidity Moisture, 4, 119 (1965).

92. H. E. Patten, Heat Transference in Soils, U. S. Dept. Agr. Div., Soils Bull., 59, 3-54.

93. B. T. Shaw and L. D. Baver, Soil Sci. Soc. Amer. Proc., 4, 78-83.

94. M. S. Kersten, The Thermal Conductivity of Soils, Highway Res. Board Proc., 28, 391-409.

95. W. M. Aldous, W. L. Lawton, and R. C. Mainfort, The Measurement of Soil Moisture by Heat Diffusion, US Civil Aeronautics Adm. Tech. Development Report 165, 1952, p. 17.

REFERENCES

96. J. Sivadjian, Soil Sci., 83, 10-12 (1957).

97. R. K. Ghosh and K. K. Khanna, Measurement of Moisture Differential in Concrete Pavement Slabs, Road Research Paper No. 57, Central Road Research Institute, New Delhi, 1964, pp. 1-20.

98. G. Szuk, Rilem 124, June 1962.

99. G. P. De Loor, Some Dielectric Measurements on Portland Cement Paste, Proc. of the XIth Colloque Ampere on Magnetic and Electrical Resonance of Relaxation, Bindhoven, 1962.

100. M. F. Legutin, A. J. Levcsinszkij, O. A. Mesedlov-Petraszjan, and G. A. Szalop, International Symposium on Moisture Measurement, Hungary, 1971, p. 178.

101. S. L. Penfield, Amer. J. Sci., 3rd Ser., 48, 31 (1894).

102. L. Shapiro and W. W. Brannock, Anal. Chem., 27, 560 (1955).